CONFESSIONS OF AN ALIEN

BOOK ONE
A MYTHOLOGY FOR THE THIRD MILLENNIUM

E.T. MARSHALL

Red Coaster Press
MILL VALLEY, CALIFORNIA

Red Coaster Press
PO Box 1523
Mill Valley, CA 94942
mythologytm.com

Editing, cover design and interior layout by Ruth Schwartz, aka The Wonderlady

Ordering Information:
Quantity sales. Special discounts are available on quantity purchases by corporations, associations, and others. For details, contact the "Special Sales Department" at the address above.

Confessions of an Alien / E.T. Marshall. -- 1st ed.
Book One of *A Mythology for the Third Millennium*
ISBN 978-0-9984272-0-1 - paperback edition
ISBN 978-0-9984272-1-8 - ebook edition
Library of Congress Control Number: 2016961634

This book is dedicated to my muse:

Lily

Also known as Mary Julia Roberts Shea

New ideas pass through three periods:
1) It can't be done.
2) It probably can be done, but it's not worth doing.
3) I knew it was a good idea all along!

—Arthur C. Clarke

It's not what you look at that matters, it's what you see.

—Henry David Thoreau

Everything is theoretically impossible, until it is done.

—Robert A. Heinlein

A NOTE FROM THE AUTHOR

Of course the title of this series is Mythology for the Third Millennium, but what is mythology? Mythology is a narrative that usually contains hidden meanings and deals with questions such as: where do we come from, what is our purpose and what is our place in the universe? Noted psychologist Carl Jung believed that our society's mythology was a reflection of its goals, fears and dreams.

This work has many references to film because film reflects our society's myths and as such, is an example of a simulation of reality that the audience accepts for ninety minutes in place of its own. Indian mythologist Devdutt Pattanaik thinks that unlike fantasy that is nobody's truth, and history that seeks to be everybody's truth, mythology is somebody's truth.

In this case, that somebody's truth is mine. As a philosopher, I have asked those questions, studied the great cultural and religious mythologies, and have mourned that those differing mythologies have caused harm. I challenge no existing mythology. For decades I have been an active practitioner of lucid dreaming and have been given a vision of a new mythology in that dream state.

Because this is a mythology, it is meant to be believed, at least during the reading. I encourage you to read this work with a full suspension of disbelief. I have purposefully made this work a short read. I hope you may find it so, and even come back for a slower second look. It is my hope that as a bell rings when struck, some of the chapters will resonate with you.

CHAPTER ONE

"I've changed my mind."

Vladimir Comsky's blue eyes were steady and serious, as always.

"What do you mean?" I asked. We had just finished watching *Inception*, starring Leonardo Wilhelm DiCaprio, for the fourth time. DiCaprio's character, Dominick Cobb, has lucid dreaming abilities and the rare talent called "extraction." Dom can enter another person's dream state and extract information.

Instead of answering, Vlad, using the remote control, jumped back and forth to different parts of the movie, stopping to point out the recurring numbers on room doors, taxis and locomotives. I was confused. I asked again, "Have you changed your mind and now think that the entire movie isn't a dream?"

"No, I continue to think the entire movie *is* a dream, a total piece of fiction in Dom's mind. That's not what I changed my mind about." Vlad paused for emphasis. "I want you to write your own lucid dreaming story."

I frowned. "Nobody's going to believe my story."

"That's exactly my point. Haven't you heard of suspension of disbelief? Didn't Steven King say something about books being portable dreams? Just write what you remember, the way it was, the way it seemed to you. You know it's been done so many times before." He smiled. If the Mona Lisa married Buddha, their child would have Vlad's smile. It was all-knowing, peaceful, eternally patient, and gave me the feeling that he was trying very hard not to make me feel stupid. His smile was very special because he doesn't have a heck of a lot more expressions. He continued, "Let me remind you of movies we have recently watched. Flying saucers landing in city parks, a blue police box flying through time and space, little green men and larger gray men. People believe those fictions. Why wouldn't they give the truth a chance?"

"But I'm no writer, Vlad. I'm a technical grant proposal hack who is told to make my submissions as dry as dust. A geek. I was trained to be boring. Who would want to read something I wrote?"

"Have you been having a boring time with me, Mr. Donald Thomas?"

I ignored the question. "Why don't you write your *own* story?"

Vlad's smile disappeared and he grimaced, as if he were sucking on a lemon. "I would sound too preachy, like Michael Rennie in *The Day the Earth Stood Still*, giving his final message to the Earthlings about 'uncontrollable galactic robot policemen.' As if giving such power to a machine was the best a spaceman could imagine," he added, shaking his head.

"It's only a movie," I said.

Vlad countered, "It's the total withdrawal of all moral responsibility if robots are going to punish the evildoers. It would be like putting Arnold Schwarzenegger's Terminator in charge, or putting the vast intelligence of the Krell in the hands of Walter Pidgeon's id, or the supercomputer from *Colossus the Forbin Project,* whose first logical decision is to empty the nuclear arsenals of both East and West, while simultaneously getting rid of Earth's biggest problem: mankind." He gestured eloquently. "You know I can go on and on about my perspective, Donald. Besides, it would be difficult to communicate your human point of view and perspective if I were writing. I might be tempted to slip in some facts that would be completely incomprehensible, whereas you, with your graduate degree in biomechanics, fluency in differential equations and Laplace transforms, cannot be tempted to reveal new technology because you do not know the mechanics of my technology. You are simply the recipient."

"Sure, I get it," I said, somewhat irked. "Incomprehensible to a pea brain like me." Vlad gave me a look that said *I'll show you what I mean*, and suddenly I was a boy again, remembering a very hot and humid day at Coney Island when older friends persuaded me to take my first ride on the Cyclone roller coaster. I think it was a New York rite of passage. I was nine years old and the Cyclone looked gigantic. For several minutes I watched the screaming coaster riders, many of them waving their hands wildly above their heads, with big smiles on their faces. They seemed to be screaming with joy. I did not see anyone injured or dead so I joined the line with my friends. I barely reached the minimum acceptable height for admission.

In those days New Yorkers came to Coney Island to escape the sweltering heat of late August. It was cooler that close to the ocean. Several Cyclone trains passed before it was our turn to board. I watched people's expressions as they exited the ride. Nobody looked too upset. Then, just before we boarded, I noticed a large overweight man wearing a white shirt and black pants hurriedly leaving the exit gate. People behind him moved slowly, wearing annoyed expressions, wiping their clothes. Apparently, the man in the white shirt had become ill during the ride and turned his head sideways to relieve the contents of his stomach, which then streamed to the riders behind him. Three workers from the Cyclone hurriedly came out and cleaned the cars of the more obvious evidence of the accident with Lysol spray and rags. By the time I got there, the smell of Lysol was everywhere. My friends honored me with the front seat in the first car, which was the seat that had no Lysol treatment because the man in the white shirt had occupied it. The three cleaners disappeared and the train slowly went up the ramp: clack, clack, clack, clack, the rhythm was steady and went on for a long time. At one point all I could see was sky. My fists were rigid around the safety bar pinning my lap to the seat. It was metal and cold. I knew that I would not remove my hands, none of this holding them above my head for me. I allowed myself the option of screaming— but not with pleasure. I expected when I reached the top that I would fall over the edge, but that was not the case. The first car went over the top, started to descend, and seemed to hang there for a moment while the rear of the train caught up.

I was suspended in the air, looking straight down, ready to drop. I had never seen the world like that—tiny and so distant, as though I had no part in it. Then it happened. The train fell in a vertical drop, the air rushed by; cooling my sweating face and neck and I felt myself relax. I loosened the death grip on the safety bar and did not scream.

The memory ended. I unclenched my fists. Vlad was still smiling at me. I looked at him, understanding that he had provoked the memory to illustrate his point.

"So how do you feel about writing what you understand," he said, "what you believe? The way it was from your point of view. Just tell it like it is. What happened and how you felt about it. You can do it. You can be a human, can't you?" I studied Vlad's face as if for the first time. The close-cropped hair graying at the temples, the hollow cheeks, the calm, penetrating eyes, the deceptively human countenance. I doubted that anyone would take him as a former resident of another planet, an extraterrestrial, the real deal, sitting here in flesh and blood in front of me. Actually, sometimes when I think about it I start to shake. If they X-rayed him, did DNA studies or any kind of chemical analysis, he would pass for a normal human. Our technology is too primitive to detect differences between Vladimir's body and ours. After all, it has been less than 100 years since we discovered the existence of DNA and its double helix construction. Nevertheless, my friend was an alien from a far distant planet. Vladimir remains an enigma to me.

CHAPTER TWO

In 1978 *Scientific American* magazine published an article announcing that the mathematician A.J.W. Duijvestijn, using a computer, had found a square which contained the least possible number of squares, 21, inside of it, of different sizes. Soon after that mathematical milestone, a letter was sent to the editor containing a photograph of a painting commemorating that event. Thirty-six thousand years, three months, fifteen days, sixteen hours and twenty-four minutes before that *Scientific American* letter was opened, an artist was completing a painting of animals on a cave wall located in what is now France. The exact time and place of the cave drawings is known to me because of Vlad. A year later, almost to the day, another man drew two perfect circles in that cave, one of the sun, the other of the moon. Unfortunately, the circles did not survive. I also know about the circles because of Vlad.

At the time of the square painting, I was working in Los Angeles for the University of Southern California, writing grant applications for their medical department to be submitted to the federal government, specifically DARPA, the Defense Advanced Research Projects Agency. DARPA's job is to prevent a strategic surprise from negatively affecting our national security by maintaining the technological superiority of the U.S. military. Luckily, the same agency also doled out monies for medical research to help wounded veterans recuperate. That's where I came in. The doctors were interested in doing their research, trying to aid wounded veterans. The veterans, of course, were interested in being healed. I was interested in helping the veterans and getting my paycheck. My technical background as a biomechanics engineer gave me some insight into the interface between doctors and the government officials doling out the money from DARPA.

The DARPA agency was created because of the shock the United States went into when the Soviet Union put a satellite, Sputnik, into orbit around Earth in 1957. It flew high over the skies of the United States sending out its message of Communist superiority: *Beep, Beep, Beep*. The government and the citizens of the United States, with the proper equipment, listened to the intimidating beeps and millions watched the satellite twinkle its way across the sky. The U.S. Defense Department vowed never again to be caught with its pants down.

DARPA's role was to consider innovative technologies in many fields that can be used to threaten America. To some, the Sputnik satellite was the start of the race into space. In the years that followed, the first walk on the moon took place, we sent Voyager probes outside of the solar system, sent cosmonauts and a space station into earth orbit and launched the Hubble telescope.

CHAPTER THREE

DARPA was not unusual as a bureaucratic federal agency in its grant requirements. It usually took me a few days to read and understand them, and then I would pull out key phrases and put them into an outline form, like a checklist. With this in hand, I would sit down with the doctors and go down the list to see if I could unite with DARPA's requirements. After this initial meeting, I would sleep on the docs' comments, and by morning usually wake to find myself writing down on my bedside notepad a list of phrases and concepts that I believed would ignite DARPA's imagination.

In one of these grant-writing exercises, the doctors were interested in helping injured war veterans whose limbs, specifically arms, had been lost in battle. They were trying to explore ways of connecting directly to the brain with an artificial prosthetic that would be controlled directly by the brain so that hand and arm movements could be coordinated electronically.

I spent hours in the basement of the Los Angeles General Medical Hospital and at the USC offices nearby in East LA. The area was populated by a predominantly Mexican community. I was in the basement because that's where my secure computer terminal was located. It gave me access to the lab's encrypted data files.

The air-conditioning wasn't for me, it was for the big computer, but I didn't mind; I was happy to be working in an air-conditioned environment. On one of those hot August nights I was working late, hoping that the outside temperature would drop to the low 90s, or if I was lucky, the 80s, before I had to venture out into real air. About nine, a janitor entered with a broom and a scoop and started to sweep around the computer area. He was wearing working clothes, a beige shirt, pants and boots. At first, I took little notice of this man as he swept slowly and methodically in the far corners of the room. I'd finished my work and was sitting there bored, waiting for the temperature outside to drop. I watched him. He was in his 50s, with unusually good posture, dignified but relaxed. After a few minutes, he put his broom aside and sat down at a vacant computer terminal. For a time he just sat there, looking around the room occasionally. In the quiet, I could hear the noise his fingers made as he ran them over the keyboard. I thought how odd this was, but since all the terminals except mine were shut down the janitor could do no harm. He caught my eye, smiled, slowly got up, and resumed his methodical sweeping. At least another five minutes had passed when he sat down at a second vacant terminal. More minutes passed and he cleared his throat and quietly asked, "Would you mind if I stayed here a while longer? It's awfully hot out there."

"Of course not," I said. How could I object? We were there for the same reason. At least I thought we were. I looked at him again and felt that he was out of place: his demeanor, posture and voice were too polished to be a janitor's. My gut told me that he might be a stage performer. We were close to Hollywood, and perhaps custodial work was his day job.

We left the basement at the same time, about eleven, parting amiably outside. It was cooler and there was a smell of honeysuckle in the air. Before I went to sleep that night I thought of the odd man with the broom and of his unusual posture. I didn't know it then, but I had just made first contact with an alien.

CHAPTER FOUR

A half-hour before our USC medical team was to meet with the DARPA team, we gathered for coffee and a briefing. I arrived, along with a few of the doctors, before our leader, Dr. Pebbles. We had the morning newspapers, and for a while the only sounds were sipping and the rustling of newsprint. Comments were voiced occasionally, to no one in particular about the current heat wave, the mayor's task force to study downtown traffic congestion, the current losing streak of the Dodgers. Mary Shaw, one of the newer doctors, recently transferred from Harvard medical school, said, "Did you see this series on homicides? The newspapers are now calling him the L.A. Strangler. Looks like you have your own version of the Boston Strangler out here."

Dr. Andrea Gupta put the sports page down and said, "Yes, unfortunately." Down the table, Dr. Henry Swartz asked if there wasn't a statistical correlation between high temperature and a high homicide rate. Just then, Pebbles walked in and newspapers and cups were put away.

"I'm glad you all made it this morning," Dr. Pebbles began. "You know how important this grant is to us and our continuing research. Just for this morning, remember to be on your best behavior."

Under her breath, Mary quipped, "Yes, Daddy." Pebbles looked at her sternly and everyone laughed. Minutes later the DARPA team entered. To my astonishment, the custodian from the basement room was with them. A moment later he was seated opposite me at the conference table.

The grant proposal was to fund research needed to effectively connect combat veteran's damaged limbs to prosthetics being developed by our University of Southern California medical team. The DARPA team's main concern was who would perform the surgical procedures. After a little bit of haggling, the USC doctors agreed to allow DARPA to use their own surgeon, Dr. Vladimir Comsky. I recognized him at once as the man with the broom.

CHAPTER FIVE

As the meeting ended, Vladimir Comsky walked over to me. "How are you, Mr. Thomas?"

"Just fine, Dr. Comsky," I responded. "But may I ask, what was all that about in the computer room with the broom?"

"I had just come in from the airport and wanted to get out of the heat and see some of your facilities, sort of to get the feel of the place. I saw you in the computer office and did not want to disturb you. So when I saw the broom I thought you might think I was the janitor, and ignore me, and go on with your work. Which is what happened." He smiled.

It was plausible, and I relaxed. He was a thinly built, wiry man, about five foot eight, a few years older than he'd appeared in the basement, around 55, with a short, almost crew-cut that poked up a bit in front. The skin on the right side of his face showed signs of surgery, including a thick vertical white scar on his upper lip just below his nose. The lobe of his right ear was missing. His hands were scrupulously clean, manicured and showed signs of scarring. He asked if I wanted to join him on his rounds. Since I had no plans for that afternoon and was intrigued by his offer, I accepted.

As we left the medical offices a large, fit man whom Comsky introduced as Mike, greeted us at the curb. When we shook hands, I felt a grip and strength that could easily have lifted me off the ground.

"Do you have the kit?" Comsky asked.

"Yes sir, Colonel." Comsky raised an eyebrow and Mike quickly corrected himself with a softer "Yes, Vlad."

"You can call me Vlad, too," Comsky said, turning to me.

Mike opened the back door of a dusty grey Chevy Impala and we got in. I expected to go to the County General Hospital, but instead, Mike drove us to the barrio area nearby. When we arrived, Vlad and I walked into the neighborhood and Mike followed at a respectful distance, keeping us constantly in sight, like a mother hen watching her chicks. Under his watchful eye I felt safe and at ease.

Dr. Comsky headed for the back alleys of the barrio, apparently looking for drug users. He approached three men sitting in the shade on soiled sleeping bags. The area had the smell of old garbage, a backed-up sewer and fresh urine. I soon realized Comsky was there to exchange dirty needles for clean ones. He would ask junkies how they were, in what seemed to be colloquial East LA slang. That much I could tell from my high school Spanish, though the words were harsher and everyone spoke faster than I could easily follow. He asked about their health and whether they had any infections, or if they were just hurting, and of course, if they wanted to exchange needles. If they showed any sign of cuts or bruises, he would clean their wounds and apply antiseptic, bandage them, and leave them with clean needles. He also left each of them a card with the address of a free health clinic in the area.

When it seemed that we were about to leave, a last group of men pointed to an even darker part of the alley. We walked further in, I reluctantly, Comsky without hesitation. He looked around, then pushed flattened cardboard boxes out of the way and found a man passed out beneath. He squatted and started speaking in another language, not Spanish, but something I'd never heard. The man opened his eyes and slowly rolled over to face him. He replied in what I would later find to be an indigenous language of Guatemala—K'iché. The man's eyes lit up when Comsky responded in kind.

The man's name was Juan Carlos Menchu. I learned later that he had recently come to Los Angeles through Mexico. He was looking for his family in the San Joaquin Valley. His money was taken from him and he had been beaten. When the street people found him, he was given street meds for his pain.

Comsky asked questions about his condition and when Menchu pointed to his left foot, the doctor carefully removed his shoe. The smell of putrefaction overwhelmed me, and I had to fight to keep from vomiting, but the doctor didn't miss a beat. He removed the sock and cleaned the near-gangrenous foot, applying antiseptic and bandaging. He also gave Menchu an injection of penicillin and a small bottle of antibiotics, a cheese sandwich and a card for the local clinic. While he was doing this, the Guatemalan said he was a teacher of history in his native country.

We wandered back to Mike and the dusty Impala. I told Comsky that I was impressed with his language skills, bedside manner and huevos for entering this dangerous side of East LA. He said someone had to take care of these people and then he looked at me for a moment and asked, "Are you taking care of anyone?"

I thought before answering. I had just met the man, and already felt connected and slightly vulnerable. "I'm sort of taking care of my neighbor, Ann," I said. "She has Parkinson's disease and lives in an adjacent house."

Without a moment's hesitation he asked, "Would you like me to look at her?" Since it was one of those summer evenings and I knew

Ann would still be up, perhaps even keeping some food warm for me, I agreed.

When I entered Ann's side of our single-story Glendale duplex, I told her that we had company and introduced Vlad and Mike. Ann and I had met twenty years before, when she and her husband, Richard, bought the other side of the duplex. Richard worked in the deli department of the local Safeway. They lived there quietly together until his death from diabetes ten years ago. Ann loved to cook the items that Richard brought fresh from Safeway and often would invite me to join them for dinner. Born in Milan, Italy, she was proud of her heritage and cooking abilities. She was in her late 70s, five foot three and pleasantly plump. Ann wore a flower-patterned, loose fitting housedress, a black pullover sweater with sleeves pushed up towards the elbows, well-worn, gray, open-toed terrycloth slippers and knee-high nylons that ended just above her thin calves. She had been watching television in a soft chair with her large, skinny, mottled orange, blind cat asleep on her lap. The cat had wormed its way into her life just after Richard's death.

She muted the television. Her Parkinson's gave her no rest, and both hands shook violently. Her head bobbed as if stuck in a paint-mixing machine.

"This is my friend and new work associate Dr. Comsky. He has experience with Parkinson's and may be able to help you."

"Y y y es, I I I would like that." She slowly got up.

Comsky asked, "What kind of treatment are you getting?" He listened patiently while she stammered on for several minutes. She had first noticed the tremors about six years ago. They would start and stop at different times. I went to a chest of drawers, pulled out her medical files, and handed them to Comsky. He looked through them while he listened to her halting narrative. When she was finished, he asked, "Do you know the drugs they're giving you for your condition?"

"They have big long names. They're over there on the cabinet," she said.

Arrayed on the top of the cabinet were a dozen black-framed photographs of smiling children and various adults from Ann's family, in addition to the drugs. Comsky examined the bottles. "Carbidopa, Levodopa. Are they working for you?"

"Yes, they t-tried a whole lot of other things that didn't work."

"Like Tolcapone?"

"Yes, that sounds f-familiar."

"With your permission I may be able to help you," Comsky said. He explained to Ann that he had experience operating on Parkinson's patients in Boston and that his methods involved brain surgery and the use of a robotic apparatus. He proceeded in giving a very interesting and clear description of his research group's activities, helping returning veterans restore use of their damaged bodies. He explained this so well that I leaned close, though I was hearing things I already knew a great deal about.

Ann asked him questions and he seemed to be delighted with her inquiries. Perhaps it was just as simple as being with a man of intelligence who demonstrated a sincere interest in her. I was content to watch her become animated. She listened intently as he spoke of his successes. His manner was calm and reassuring. Before he left, he asked Ann whether he could contact her doctors about her condition. She agreed. Would she allow him to help her with her treatment?

In a small voice that came out like a stutter, she said, "Yes, yes, p-p-please. Please help me."

After Comsky left, Ann said, "I-I-I like this doctor, I trust this man. What do you think, Donald?"

"He seems to have a lot of good experience and credentials and is working with the best hospitals. I like him too."

CHAPTER SIX

The next morning Mike came by my home, handed me a manila envelope and said, "He asked me to personally hand this to you."

"What is it?"

"He said you would know."

I opened the envelope to find a stapled DARPA grant submittal application about an eighth of an inch thick. Apparently, Vlad had taken the liberty of distilling the DARPA-USC meeting contents into a complete grant application that I presumed would be entirely acceptable to the grantors. Of course it included the provision that Vladimir Comsky would be the designated surgeon on the USC medical team.

After reading the proposal, I asked one of the office clerks to photocopy the grant application and place a copy in each of the in-baskets of the doctors who had attended the DARPA meeting. I returned to my office to look at my empty in-basket, checked my secure CPU for emails, and then strolled down the block to Starbucks. When I returned a couple of hours later, the chief medical officer called me into his office and asked me to sit down.

Harold Pebbles was a tall balding man of 60 with a boyish face. He was the son and grandson of doctors. His great-grandfather, a medical officer for the state of Georgia in the War of Northern Aggression, had moved to Kentucky after he became a doctor, maintaining a small country practice. Upon the old man's retirement, Harold's father took over the practice and sent his son to the Perelman School of Medicine in Philadelphia, the City of Brotherly Love. There, Harold met Teresa Dennison, who was studying at the Curtis Institute of Music, obtaining her Master's degree with a focus on piano. Teresa's family raised thoroughbred racehorses.

Teresa and Harold soon found that both had brothers injured in the Vietnam War. Her brother, William, had his left arm amputated, while Harold's brother, Timothy, lost his right leg below the knee. Harold decided to devote his life to helping disabled veterans. He had married well; there were photos of his wife's racehorses on the walls of his office. When he was thinking or listening intently, he would rub his simple gold wedding ring with his thumb, during which the top side of the ring seemed to move by magic. He was doing that now. I waited to hear why he'd summoned me, but he only shuffled papers around on his desk. Finally, he picked up his copy of the DARPA governmental military grant application, pushed the button on his intercom, and said, "Please ask them to come in now."

The door opened. The doctors filed in, each holding a copy of the application, and sat down quietly. No small talk about the traffic and the weather now. I waited.

Pebbles addressed me. "You have been a busy boy, Donald. Have you been up all night writing this? And what about the research details?" His tone hardened. "Did you access all of our doctors' research files?"

"But, sir, you know all this information is online, on our internal servers and of course, it is all encrypted and secure," I said meekly.

"Yes, of course I know all that, but this work, this content, the detail? Doesn't it usually take you weeks to come up with even a draft, and then longer still for it to be reviewed by the people in this room? But this morning the people in this room reviewed this application and found it not only to be thorough, detailed and correct, but also asking for more money than we had originally budgeted."

"I thought we could use a cushion," I improvised awkwardly. I knew that the application was a slam-dunk provided by Dr. Comsky, and that he wanted to be sure that he would be named as the operating surgeon. I had mixed feelings about taking all the credit but concluded that because Dr. Comsky had not put his name on the application, it was his intention that it should appear to come solely from the USC medical team. I wasn't sure why, but by now I trusted the man. So I added, "I believe we should expedite this project, Dr. Pebbles."

"Yes, of course. Good work, Donald." Pebbles smiled, and I saw the gold ring moving.

On my way back to the office, my cell phone rang. The number was from an unfamiliar area code. I was still in a good mood from the meeting, so I answered.

"Mr. Thomas? It's Mike, Dr. Comsky's assistant."

"Hi Mike, what's up?" I figured he wanted to know the outcome of the meeting.

"We need to talk. Can you meet me for a beer?"

I accepted. I felt my work for the doctors was done for a while, and now I had the possibility of finding out more about Dr. Comsky.

CHAPTER SEVEN

A half-hour later we were seated on red vinyl benches at a yellow Formica table, next to a large mural of an Aztec warrior holding a young woman in his arms. The room was lit by sparse florescent lighting, ceiling fans rotated slowly, and a yellow sticky tape hung in a far corner, advertising a dozen successful kills. Mike ordered Pacifico and I asked for a Heineken and an order of nachos, my favorite—tortilla chips smothered in melted cheddar cheese, ground beef, guacamole, sour cream and chili sauce.

"Thanks for meeting with me, Mr. Thomas."

"Please, Mike, call me Don. Why the urgency?"

"No real urgency, Don, but I thought you should know that Dr. Comsky is doing a few things that are unusual even for him." Mike drank his Pacifico straight from the bottle. I ate a chip with cheese, still hot and soft, and watched him fidget. He looked to be in his early 40s, short-cropped brown hair and dark eyebrows over very dark brown eyes. His profile reminded me of the Indian head on the face side of a nickel. He was at least six inches taller than me and 50 pounds heavier, and could easily pass for a bouncer. I poured the

Heineken into a frosted mug and drank it with another piece of warm nacho. I invited him to tell me more.

"About a year ago, he and I were stationed in Afghanistan. It was 120 degrees in the shade at the height of the day and the wind blew fine dust constantly. You breathed it in your nose. It was gritty in your eyes. When you ate your food, it was like someone added a pinch of sand. Except, of course, when we were in our secure base, there it was air conditioning, ice cream, turkey sandwiches and no sand in your mouth. Colonel Comsky asked me to drive him to a village that was just secured. He would do this often when he was off duty. I felt honored to drive a high-ranking officer; the Colonel said he felt safe with me as his driver. We started off at 11 o'clock in the morning. By the time we reached the village, the temperature was a cool 98 degrees. I parked the Humvee in the shade of a building. We started to walk around from house to house, looking for people who might need help. There was no shortage of those. I remember that it was about one o'clock, and we were on one side of an open area, about 50 yards from the Humvee. Dr. Comsky had just finished helping a family and we were leaving their house. I took a step from the shaded interior to the bright sunlight, heard a rifle shot and felt a sharp pain in my left foot and my left calf cramped. I fell back into the shaded doorway. The Colonel looked at my face and gave me a shot of morphine right away. There was a hole in the heel of my left boot. He carefully removed the boot, moved my foot around and felt my calf. Then he said, 'Sergeant Pickett, your Achilles tendon has been severed. You won't be doing much walking for a while.'

"I asked how we were going to get out of there. The shooter was still out there and the radios were in the Humvee. He asked me if I'd ever done a three-legged race when I was a kid and I said I had. I was hurting like hell. The Colonel spent the next few minutes bandaging my leg and stopping the bleeding. The family in the house formed a loose semi-circle about five feet away from us. The man of the house asked if we wanted tea. The Colonel answered that he was grateful for the offer but we were leaving soon. After a few minutes he picked me up, placed my left arm over his shoulder and around his neck. I remember being surprised at how easily he picked me up; I never thought that such a slight man could be so strong. He asked the husband to tie a cord from my left ankle to his right ankle. As we stood in the doorway, bound together, about to step into the blazing sunlight, I will never forget the words he said to me. 'The bullet entered from your left, so the shooter is on the left side. I will be on your left side, as we walk to the Humvee. We are both wearing camouflage uniforms. I want you to walk slowly with me; try to keep your right arm down by your side. Are you ready?'

"I said. 'I'm going to use you as my shield.'

"The Colonel gave me a funny smile just before he leaned over me to my left side and moved my left leg, now tied to his right one, into the sun. Then his left and my right, then his right and my left again. It was only after the 12th step that I heard the crack of a rifle shot. The Colonel kept his slow pace. Step 16, another Crack! When we were ten yards from the Humvee I wanted to make a run to the finish line. My right arm lurched out as if we could take a larger and faster step. Dr. Comsky turned his face to me when he saw my right arm lunging sharply forward. Crack! The shot echoed between the brick buildings. The left side of my face was spattered with the Colonel's blood. The right side of his face and neck were covered with blood. Another Crack! I was not hit, but I dropped my right arm limply to my side. Crack! The Colonel continued to slowly walk the last steps to the Humvee. Covered by a building now and next to the Humvee, the Colonel cut the cord binding our ankles. He helped me into the passenger side of the Humvee, and then drove like a bat out of Baghdad to our base.

"When we arrived at our base, just as I was sent to emergency triage, I heard the Colonel say, 'I'll do him.' He appeared about a half-hour later in the operating room with a large bandage on the right side of his face and a clean surgical gown. He reattached my Achilles tendon, and put a cast on my leg. And he took a bullet for me." It was only when he reached the end of his story, where he repeated Comsky's words "I'll do him," that Mike's voice cracked a little.

CHAPTER EIGHT

I had finished my beer and half the nachos, though I admit eating got harder as I pictured Comsky's blood-covered face.

"That's a great story, Mike," I said,

"Thought you should know. Want another round?"

"Sure."

Telling the story seemed to free his appetite. When the two fresh beers came, Mike poured his into a chilled mug, scooped two tablespoons of hot sauce onto the corner of the nachos near him, and quickly placed one of the chips heaped with beans and hot sauce in his mouth, followed by some cold Pacifico. "Those weeks in a cast helped me decide not to re-enlist. Then, when I heard that the Colonel was going to retire, that clinched the deal. I ran into him in a bar at LAX. We reminisced about all the good times we'd had, and the bad. He asked what I was up to. Nothing, I told him. He asked, 'Want to work for me? Four times your Army pay plus gym membership?' I laughed and said, 'Sure Colonel, and what will I be doing for all that money and gym membership?'

"'You'll be my assistant. On call 24 hours a day. Do you have a car?' He was all business. I said no. He told me, 'Buy one for us, a used gray Impala—I think that will do.' He handed me a black credit card and told me I could take my salary, gym membership and the cost of the car, and anything else we needed out of this. He said, 'I'll get you one with your name, and you can call me Vlad from now on.'"

Mike paused. "I swear that's the way it started, Don, about a year ago. I caught onto his routine, which, as you know, is going into poor neighborhoods and helping sick people, just as he did in the Army. In the beginning, I thought he was taking me on as a charity case, that he felt sorry for me, that to him I looked lost, now that I was not in the Army. Perhaps I was lost, I was trained to kill and I did my job well."

"So what changed, Mike?" I asked, popping the last of the nachos into my mouth.

Mike made an impatient gesture, "He took you on one of his needle exchange trips and for the first time asked me to stay back in the distance. He did your grant application for you. He wants to personally help your neighbor, Ann. I don't know what's going on with the Colonel, I mean Vlad, I'm his assistant, I didn't question his orders when I was in the Army and I don't intend to start now."

"I'm as puzzled as you are. Why's he doing all this?"

"I don't know the answer to that question, but I do know where I stand with the Colonel. He gives me an order; I carry it out. Even though I'm no longer in the Army, I am old school Army, and I will protect that man, even though he has never asked me to. Since he seems to trust you, so will I."

"Good," I said.

His expression changed. "You should know that I am licensed to carry concealed weapons and have no hesitation in using them. Since I have left the Army I have not used that part of my training." For a moment I felt intimidated by this well-built, large and armed-to-the-teeth ex-Army sergeant. I then relaxed at the thought that he was on my side, or more correctly, our side. I waited until he took another swallow of beer, and then asked him, "Is that it?"

"Pretty much so. I thought you should know."

So now I knew that the man with the broom led a very interesting life, and had his own personal assistant who for all intents and purposes, was a personal, very loyal and lethal bodyguard. I imagined the numerous trips that Mike had taken with his Colonel into hostile, poverty-stricken areas, doing needle exchanges and bandaging wounds. How many times had Mike been tempted to hurt, maim or kill?

CHAPTER NINE

After a few days, Vlad arranged through Mike for Ann and me to join him in Boston. After consultations with her doctors and several CAT scans, he was able to determine the area of high brain activity when her tremors occur. He outlined his course of treatment. He would probe Ann's brain to locate the faulty neurons causing her tremors. A very thin wire connected to a microchip would be embedded between her skull bone and her skin. His colleagues took measurements from the electrical wire interface during the trembling seizures, and then reprogrammed a computer chip to counteract the electrical impulses of the faulty neurons.

He was happy to discuss all aspects of the pending operation, reprogramming the microprocessors with me, although it was way above my pay grade. While she was being prepped for surgery, Vlad showed me around the operating theater. "You see, Donald," he said, pointing to a large piece of machinery next to the operating table, "this robot will actually be doing the delicate work on Ann's brain, and I will be controlling the robot. I will look through these binoculars and manipulate the robot with these gloves. I will use a small wire to monitor the firing of her nerve cells and make a microelectrode recording. I want to find out which specific areas and nerves are causing the problem so that I can place a neural stimulator in that precise area of her brain."

"Will it hurt?" I asked.

"We will use a local anesthetic to penetrate the skull. However, the brain itself has no sensory organs of its own. In fact, she will be awake for the procedure. Back in LA, we used computed tomography scanning to locate the general area where Ann's Parkinson's disease erratic nerve signals originate. Now it is simply a matter of using the neurostimulator to interfere with those signals."

Simply? I thought. It didn't sound simple to me.

They let me briefly visit Ann before her operation. She looked calm and alert and was wearing a scarf on her head, I assumed so that I would not see where they had shaved her head. She greeted me with a big smile and held out a trembling hand. As I held it I asked her, "Can you think of a happy thought before they take you to the operating room?"

"Sh-sh-sure. I'll think of you."

"Come on now, Ann, what was the happiest time of your life?"

She said after a moment, "When I was very young and Richard was courting me, b-bringing me flowers and candy and speaking s-s-s-so sweetly to me."

"That's the stuff. Think of that before you go and I will see you soon." She smiled, and was still for a moment as I left the room.

I had difficulty sleeping that night. The next morning I hurried to the hospital without even stopping at Starbucks. I found Ann in the Recovery Room, sipping orange juice through a straw. When she saw me she gave me a large smile and extended a steady hand in my direction. "It's a miracle," she said. Gone was the shaking that had plagued her for so many years. Hardly able to believe it, I sat at her bedside holding her hand, tears in my eyes. Both Ann and I agreed that Vladimir Comsky, his microchip and the whole field of medicine were miraculous.

By now, Vladimir had secured his place as a competent surgeon with my doctors at USC. He successfully restored the movement to an injured veteran's hand by surgically installing a nerve bypass between the veteran's upper arm and hand muscles. The patient had been a sniper, with two tours of duty. He was returning from a routine mission when an IED detonated under his Hummer. A small piece of shrapnel had severed the nerves at the wrist, precluding his use of his right hand. He said he was happy to be alive, since the rest of his squad hadn't made it.

Mike and I were surprised to be invited to view this operation since we weren't officially part of the operating team. I gratefully accepted, but had little understanding of what I was watching. It was only after the operation was over, when the team doctors spoke among themselves, that I began to understand the awe in which they held Vlad's artistry. They had watched a surgeon choreographing a dance with scalpel and sutures. Or to put it another way, it was as if they had come from a movie premiere with Oscar potential for the leading man. Their faces glowed with admiration.

After weeks of physical therapy and rehabilitation, the sniper would be functional. And then he would be returned to active duty, remaining a member of the United States military. I did not ask what Mike thought about this.

CHAPTER TEN

Soon after Ann's operation, Vladimir Comsky invited me to his house for wine and cheese. He had a two-bedroom, two-bath tract home in Van Nuys in the San Fernando Valley The valley could be swelteringly hot in the summer, as it was now, but inside his home it was cool and serene. The entire interior of the house was painted bright white enamel. Only the wood beams in the ceiling were left exposed. No carpets on the floor, only wood. Spartan. The living room's two reclining chairs faced a large video flat screen and a bookcase, which held, in addition to books and rocks, a component stereo system, an amplifier, preamplifier, a CD/DVD/Blu-ray player, record player for vinyl long-playing records, speakers, an antique tape recorder, and FM tuner.

"Have you ever noticed, Donald," he asked as we sat in the reclining chairs, "that I can record music from the FM to the tape recorder while listening to it at the same time?"

"Sure. I can do that with my own system."

"But have you noticed that this component system can work as a metaphor for the human nervous system?" Without waiting for an answer, Vlad rose from the recliner, walked over to the bookshelf, and pointed out the Y adapter that connected the output of the FM tuner to the preamplifier then to the tape recorder. He said, "Look at this; it is very similar to the organization of the human nervous system. What if--?" He stopped, took a deep breath, then sauntered back to his chair and sat back comfortably. He said, "What if the human nervous system operated in such a way? Do you know, Donald, about 100 years ago there was a great debate as to whether there was a continuous nervous system or a series of interconnecting cells, which are now called neurons, that communicate with each other through synapses? At that time, microscopes were not strong enough to see the gap, the synapse, between the cells. It was only when science caught up and could image individual neurons and their interconnectivity that the debates were concluded." He smiled. "Yes, indeed, the human nervous system—and in fact, all nervous systems—are made up of interconnecting individual cells, neurons. The connections are made up of synapses with chemical communications between them. That's what I have been doing for DARPA, reestablishing those chemical connections between the neurons and synapses through artificial means."

"I know this, of course," I said. "I've been watching it happen for years. And you are a master at interfacing and reestablishing those neuronal connections."

"Yes, but there's so much more. Imagine, Donald, if at those synapses, the electrical firings could be recorded just as the music could be recorded to that tape recorder and then re-inputted into the synaptic stream! That is what I've been trying to do with these war casualties—restoring the synaptic stream. Just think of it, Donald, capturing the stream, recording the stream and then playing it back. The muscles will respond to the appropriate synaptic stream. The lame could be made to walk again."

"It sounds so simple and I can see the possibility, but the execution is still out of our grasp."

Vlad leaned forward intently. “Donald, can you imagine that if the synaptic chemical and electrical communication between neurons was recorded and stored, say, in the optical nerve, and then at some later time reintroduced into that synaptic network, what the effect would be?”

I said slowly, trying to think it through, “Well, in theory, it would seem if there was no additional input to those synaptic connections, the identical stimulus to those optic nerve impulses should produce the same effect.”

“Yes, you get it.” Vlad seemed almost excited.

“So, in theory,” I said slowly, “just to fantasize a bit, if I saw something that was interesting and a device recorded the chemical interface between the synapses in the optic nerve and those same electrical impulses were then put back, inserted, replayed into those synaptic junctions, I would see the same images that I would've originally seen. Is that right?”

“That's right, but just to be clear, consider that television set in front of us. I have this remote control; I can push its button and turn it on. Consider the analogy of having such a system within your neural synapses. I activate the playback system to your neural synapses. You now have access to an artificially constructed playback system directly connected to your optic nerve. You close your eyes, but now you see what is artificially produced by the same system. Now, look at the video screen.” He pushed the power button on the remote. “You take what you see on the screen as normal. It is, in fact, an artificial stimulation of your optical nerve. It is not part of natural reality. But what if instead of the TV screen stimulating your optic nerve, your optic nerve was being stimulated by a stored stimulation at the synaptic level?”

“Well, Vladimir, this sounds like science fiction stuff.”

“But do you see it is possible?”

“Sure, as a science fiction.”

"You like science fiction, don't you, Donald?" Comsky continued softly. "You see what we've been doing at DARPA was bypassing the neuronal pathways with microcircuits to reestablish working limbs. We are inputting electrical signals so that limbs can function. You know that we've been doing this and also stimulating the optical signals through an artificial retina. We have been stimulating neurons so that deaf people can hear through cochlear implants. It's happening right now, Donald."

"Yes, I know this part is not science fiction. It's a miracle that the deaf can hear." I had been invited for wine and cheese, but so far no food or drink had shown up. How long was this discussion going to go on? Vlad seemed not to notice my discomfort; he was on a roll. And I was fascinated, too.

"So just imagine," Vlad continued, "an optical implant directly to the optical nerve and also directly to the auditory nerve. It would be just as if you were watching television visually and with sound, except that it would be right in your head. If you closed your eyes and were in a soundproof room with no distractions, it might seem just like being there."

I was taking this all in, watching him, and looking at the blank television screen. I closed my eyes, and my mind reeled with the possibilities.

After a few moments he rose, picked up the remote, and said, "Now consider this device. I can push a button and change the channel. It's instant." He switched through a few TV channels. "I can look at a live broadcast from the other side of the world. I can look at videos recorded from the past. I have a DVR and can look at movies repeatedly, stop and replay just the parts I like, all with my remote. What if that was all in your head?" He paused, but did not appear to expect an answer. Having known this skilled surgeon and been in awe of his capabilities, I could do nothing but sit and take it all in. What was the man telling me?

After a few minutes of awkward silence, Vladimir asked, "Do you want to know more?"

"Of course," I said. "But do you have anything to eat in the house? You did invite me for wine and cheese."

Vlad started. "I'm so sorry, Donald. I often forget to eat. I'll get you something right away." He disappeared into the kitchen and moments later returned with a plate of cheeses, a soft Gorgonzola Dolce Igor which tasted like a salty blue cheese, a firm Cromwell Somerdale which had bits of minced onion and chives that stuck in my teeth, a cut French baguette, Mary's super-seed everything crackers with garlic and onion, and a bowl of pesto sauce.

I began to eat and right away felt better. I dipped bread into pesto and lifted it to my mouth; it was deliciously garlicky. He then brought out a bowl of black Cerignola and green Castelvetrano olives and a chilled bottle of Marie de Beauregard Vouvray, a French Chenin Blanc. Comsky did great wine and cheese, and how did he know about my favorite olives? Savoring the first sips of the smooth white wine, I leaned forward and said, "Now, tell me all of it."

Vlad lowered his voice. "May I confess to you?"

"What do you mean?" I asked, heaping soft Gorgonzola on Mary's cracker "I'm no priest. What are you talking about? "

"I am an alien," Comsky whispered. "As simply as I can put it to you, Donald Thomas, I am from another planet."

I took a long look at Vladimir. I found myself slowly shaking my head. "Are you kidding me? You have a spaceship hiding around here somewhere?" I looked for a smile or some sort of other sign that he was putting me on. He sat there looking thoughtfully at me. I slowly chewed the remains of my cracker.

"I expect you have a lot more questions," he finally said.

I reached for another cracker. "Okay, I'll play along," I said. "Aren't you afraid that the powers that be, you know, the folks in the CIA, the men in black, in fact, our own Army would love to get ahold of your alien technology? By the way, what is your alien technology? I haven't seen it."

Vlad's expression did not change. "Are you interested in seeing a ray gun?" he said. "Did you know that the Army has a laser cannon that can blast through steel and knock a satellite out of its orbit? Maybe it would be easier if I just showed you. Have you heard of lucid dreaming?"

"Funny you should ask," I said, happy to get him off his alien kick. I took a large mouthful of wine to try and rinse the seeds from my mouth. "I have been trying to do lucid dreaming for some time."

CHAPTER ELEVEN

Some years ago, Harold Pebbles introduced me to one of his Philadelphia classmates. Rita Goldblatt and I found that we had a mutual interest in dream interpretation. Rita introduced me to the Nova Dreamer. It looks like a sleeping mask that covers your eyes. Batteries operate small sensors that detect rapid eye movements, REM, that indicate a dream state. The machine can be programmed to have its sound and lights activate when a REM state is detected. The sleeper is gently awakened, to be lucidly aware of his dream state. This machine helped me remember my dreams, which, after I had fully awakened, I wrote down on a legal tablet kept on my nightstand. The machine later facilitated my ability to create my own dreams and to be an active participant. Rita took a great interest in my dreams and their meaning over the years. I related this to Vlad, who listened keenly.

"Then you know," he said, "when you are in a dream state, it is often difficult to know that you are dreaming and not actually experiencing the real world. That is what I want you to do, before you go to sleep, simply say aloud that you would like me to come visit you in your lucid dream and show you the alien world I live in. At any time, you can ask me to awaken you, bring you back, and that you want to vividly remember your entire dream after you awaken. What do you have to lose, Donald—a good night's sleep?"

I decided to give up my efforts to get Vlad off his alien theme and go with the flow. "It seems harmless enough. Are you going to hypnotize me or give me some suggestions on what to dream about?"

"No, none of that."

"Well, Dr. Comsky, I am comfortable right here in this lounge chair. What if I just leaned back, put my feet up and dozed off right here?"

"That would be fine, just say the things I've asked you to and off you go," Vlad said reassuringly.

I thought this might be something good, or at least new. There was not a thought in my head that Vladimir would do anything to injure me, nothing to even upset me. I believed that he would have some sort of suggestive power to influence my dream state. I was definitely up for something new. I felt a little sporty, as if I were a trout being pulled in on a very light line. Whatever was on the other side of the line, I decided to allow myself to be reeled in. I leaned back in the chair, kicked off my shoes, loosened my belt buckle two notches and unbuttoned my top shirt button." I murmured, "I want Vlad to visit me in my lucid dreaming state. I want him to bring me back at my request. I want to remember this lucid dream vividly when I awaken." I added as a joke, "I want him to show me his planet." With that I closed my eyes, let out a sigh and repeated my last words silently, and could not help think of the White Witch from the Wizard of Oz saying to Dorothy, "Close your eyes, tap your heels together three times, and think to yourself, *There's no place like home.*"

CLICK

CHAPTER TWELVE

When I opened my eyes Vlad was still seated in the chair next to me. "What happened?" I murmured. "I don't remember a thing."

"Oh ye of little faith." He smiled as he said this; I looked around the room. The walls were disappearing. It was dark out and the sky was full of stars. I felt a chill and started to shake a little. Then, as if someone had once more pushed a button on a remote control and changed channels, the stars in the sky were different, and there was no ground below us. We were still in our chairs and the chairs were still on the floor, but that's all there was. I was wondering how we could breathe, but I remembered *this is a dream and I am letting Vlad control it.* Mercifully, the wine and cheese were still between us. I popped a large black olive in my mouth. So far, the dream was damned good.

Looking at the black sky around me and the pinpoints of the stars that were both above and below, I felt like I was in a planetarium. I sat up from my reclining position and asked, "Why did you bring me here?"

"I thought you might like to see another planet," he said, pointing his thumb behind him, as a hitchhiker would. I turned to look, and the chairs and floor, or what was left of it, swiveled around so that we were gazing down on the top of an enormous blue-and-white sphere. The view reminded me of Earth as seen from the space station. Slowly the view changed so that I was looking straight at the planet and no longer down on it. Like my experience on the roller coaster, this made me feel as if any moment I would fall out of my seat and down to my death on the planet's surface. I could see large landmasses and oceans, but knew that this planet was considerably larger than Earth and had entirely different-shaped continents. I finally said, "It's awesome."

"Yes, I thought you'd like it. Its name is Paucri. Can you see it clearly? And do the stars look like pinpoints? And does my voice sound natural?"

"Yes, of course, why?" I asked.

"You know this is a simulation. It is not real. But does it seem real to you?"

"Indeed it does." Except for the disappearing walls, it seemed totally real. I sat there for a while trying to take it all in: Paucri, the stars, my alien friend and the non-reality of this reality. I thought of a chimpanzee looking at a television set. I thought of an alien looking at a chimpanzee. I thought of an alien looking at me. I don't know if it was the height, the new experience, or the chimpanzee, but I felt strange and uncomfortable. .

Vlad, sensing my discomfort, asked, "Would you like to go someplace else? Somewhere more familiar to you?" Gratefully I assented, then added, anxious not to give the impression that I didn't trust him, "What do you have in mind, Vlad?"

"Where would you like to be right now if you could pick anywhere on Earth, at any time, and while we are at it, you can pick the weather, too."

I thought for a few moments of the pleasant times of my life. The movie City of Angels, with its vivid sunset scenes, flashed into my mind. More than once at sunset time I would find myself at a beach—Malibu, Santa Monica, Huntington, Redondo—but it was Venice Beach that called to me now. A place of warm memories of ice cream cones, hot dogs, cotton candy, roller skating and watching the sun go down with the woman I loved. "How about Venice Beach today an hour before sunset and 80 degrees?" He nodded and somewhere, a button was pushed on a remote control.

CLICK

CHAPTER THIRTEEN

We were sitting on beach chairs facing the Pacific Ocean. There were many clouds in the sky but no sign of rain, and it had all the makings of a beautiful sunset. I watched the bright reds, reflected on the clouds above the glow, softly and slowly transform into a deep purple. A dozen squawking seagulls were flying overhead; sandpipers were darting in and out of the surf line. I took my shoes and socks off and rubbed my feet in the warm sand. I thought how easily my mind could be fooled. One minute in space and the next at Venice Beach. As I looked around, it sounded and felt so real. I could hear the waves breaking gently on the beach, the tender wind moving through the palm trees behind me. Vlad refilled my wine glass.

"But Vlad, where are the people?" I asked. "Venice Beach at this time of day is full of people."

CLICK

As if a thunderclap blasted behind my head, my body uncontrollably jumped back into the beach chair. In an instant, the sound of hundreds of people surrounded me, as if someone had turned on the speakers of a stereo system. All the people were moving as if they had always been there. I looked around from my beach chair at the sidewalk behind me and saw dozens walking, a few roller skaters gliding between them, and of course, cyclists. There were children playing in the sand on the beach, people eating pizza and hot dogs, and lovers sitting on blankets looking dreamily in each other's eyes, arms wrapped around each other's shoulders. Then I noticed the smells of those hot dogs, and somewhere down the beach, someone was barbecuing pork ribs and hamburgers, and everywhere there was the smell of suntan lotion. I inhaled deeply, trying to see if this was real or not. I put my hand in the warm sand, rubbed the fine grains between my fingers, and sniffed it.

The simulation was outstanding, and the thought shook me. What if we went back to Van Nuys, and it's actually a simulation? I believed I wouldn't know the difference now that I could feel the sand on my fingers. But Vlad did promise to take me back at my request, and I did believe him. After all, he was the man who cured Ann. I trusted him then, I had faith in him, so why should that change now? I questioned my own judgment and reality. I admitted to myself that I was afraid.

CHAPTER FOURTEEN

Why had Vladimir chosen me? Was I so very suggestible? What he was doing was so powerful. I said, "You should share all this with the government, the United Nations, the president of the United States!"

"You mean share this dream world?" Vlad said. "How do people treat dreamers, people with their heads in the clouds?"

I countered, "I remember a man who had a dream and made a speech about it in Washington DC. Thousands of people remember that dream. Many of them changed their lives for the better because of it."

"That's true. Donald, but unfortunately that's the exception that proves the rule."

I had no answer. I was sitting at Venice Beach watching the sunset in an unreal world with an alien from another planet, thinking how ugly reality is, thinking of the violence, thinking of the hatred, thinking that for right now, I did not want to return. I sat there quietly taking it all in. Mothers carrying their children and their pails away as the air turned cool, lovers wrapping a blanket around them and snuggling even closer. We sat there until the sun went below the horizon and the sky turned purple and orange. I watched the contrails of a jet overhead and wondered if it was military. My mind wondered to a Star Trek episode where the Enterprise goes back in time and is pursued by a US Air Force jet. Of course, the military is interested in aliens.

I now fully believed Vlad was what he said he was. I looked at him in wonder. "Sometimes I forget who, or what I'm talking to. You must have your reasons for not talking to the president of the United States, don't you? Care to share?"

He didn't answer. "Are you getting a little cold, Donald?"

"Well, now that you mention it, yes, and how are we doing for time?"

"First things first," Vlad said. "I will bring us back to the same scene an hour before sunset, the same seagulls, the same sounds, the same clouds and even the same sandpipers." The remote button was pushed again.

CLICK

"Vlad, you are good," I said, looking wonderingly around at the empty beach. "And my question?"

"Was that about how we're doing for time? Are you worried that because we are spending so much time here that you may be late for work tomorrow? Or that you may not get enough sleep tonight?"

"Something like that," I said, reluctant to say more. I had second thoughts about asking about time. I wasn't sure what time meant in this non-reality. I wasn't sure what time meant to Vlad in this simulated world. Was it like a dream, where days and weeks go by, all seemingly in the blink of an eye? I watched Vlad take a deep breath, preparing to answer, and to take him off the hook I said, "Forget I asked. I'd like to hear about why you're not talking to the president, if you don't mind." I saw him relax, seemingly relieved that the question he would have to answer was so simple.

CHAPTER FIFTEEN

"Have you heard of the Rio scale, Donald?" Vlad asked.

"No."

"It was proposed by the International Academy of Astronautics. A group of your top scientists got together in Rio de Janeiro and decided that there was a low probability of discovering an alien civilization. Even though the probability was extremely low, if it ever happened they believed it would upset the world deeply. These scientists, with their analytical and highly respected minds, broke down the impacts of an alien contact along the lines of a few simple questions: how far away the aliens seem to be, how the aliens present themselves, what type of aliens they seem to be—friendly, hostile—and how credible or believable is the contact with these aliens. Well, Donald, here I am. What you think? How far away do you feel I am? How real do you think I am? And what do you think of me as an alien? I have to see where I fit in on the Rio scale with you."

"I think I can answer two of the three questions fairly easily," I said. "You are right here and I think you're very real, at least I plan to think that when I get back to reality and off this beach. As for you as an alien, although you seem very human to me, and a gifted surgeon, except for this business of a dreamlike simulation, which, by the way is just great. In addition, as far as this dreamlike simulation goes, I'm not entirely convinced that this is not the result of some sort of hypnotic effect. If that were true, then you would be a non-alien who is also a gifted hypnotist. If that were the case, I have to ask myself why would you lie to me? What would you have to gain? Would I like you any less if you were simply a gifted hypnotist? This beach illusion, the seagulls, the sunset, the stars, and Paucri are all fabulous illusions, which I am enjoying enormously. So I am having difficulty answering the last question of the Rio Scale." I looked at him anxiously. "Is that a fair answer?"

Vlad seemed to relax in his chair; I could see this even while I was speaking. His voice became a notch lower and his words emerged more relaxed and slower. "Donald Thomas, I enjoyed your answer enormously, but to continue answering your question about why I just do not walk up and talk to the president of the United States, let me explain. The Rio scale is a logarithmic scale, similar to the Richter scale for earthquakes. The Rio scale goes to 10, zero means of no importance, six being a noteworthy contact with an alien—sort of lukewarm—nine being of outstanding importance, and the top of the scale can be described as of extraordinary importance. So, Donald, you can choose to understand me as a delusional, gifted hypnotist trying to trick you into believing that I am an alien. Just for a moment, consider yourself deluded, that you actually believe I am one. Consider also that I walk up to the president, assuming I can get through all of his security, and say, 'I'm an alien here to help you do great things.' The top scientists who developed the Rio scale have concluded that this would be a 10 on the Rio scale, an event of extraordinary importance, an event that would change the course of Earth history, and an event that would have deep social, political, religious, intellectual and scientific consequences. This last part I am quoting almost verbatim from the people who developed the Rio scale. Donald, does this sound like a good thing?"

"I see. So you don't want to come out of the closet because you think there will be bad consequences."

Vlad said, "I have to agree with the great minds that produced the Rio scale. Exposure to an alien may cause rioting in the streets, a stock market crash or general mayhem."

I stared at Vlad as he was agreeing with Earth's "greatest scientific minds," then looked up at the clouds turning dark and silver with the sunset, listened to waves gently breaking on the beach and to the seagulls, now only occasionally giving out a cry. I thought how wonderful it was to be deluded by this skilled hypnotist/alien. Did it really matter which he was? If Vlad really thought himself an alien, why should I disbelieve him? I've been told that great conjecture requires great proof, or words to that effect. So where was the proof that Vlad was an alien? I'd seen Paucri. Was it all an illusion? As I sat there enjoying the sunset I asked myself again, does it matter? Perhaps I should just wait and the proof will reveal itself. In the meanwhile, whatever Vlad was doing to create these marvelous illusions was damned good. And if it was only that Vladimir Comsky was a great human surgeon with a vivid imagination who had performed a seeming miracle on my friend Ann—that was fine too. I said, "Okay, Vlad, I get that you don't want to upset people with alien contact, but can I ask you another question?"

"Go for it."

"If you're an alien, why are you a surgeon working for DARPA?"

"The simple answer is DARPA is where the action is," he said. "It exists in the most technically advanced nation on the planet, in the most technically advanced military organization, and it has a role to seek out and explore the leading-edge of technical developments. But that doesn't answer the question of why I'm here as a surgeon. Many of the technical advances presently considered by DARPA have to do with computer neuron brain interfaces. There are many ways that things can go wrong between neurons and computers. You know about this with your biotech engineering background. Among other things, I am intensely curious about how your government and DARPA are making decisions regarding this interface and how they treat their veterans. The reason I have arranged to do the operations myself is that I would like to eliminate, as much as possible, human error that may introduce itself into the surgical procedure. I am not interested in doing anything new. I am interested in doing—very well— already established procedures. I hope you don't mind my humor if I say that I want to eliminate human error."

I liked his pun, and said so. "I understand. You're here to help."

"I don't mind your looking at it that way."

I was not happy with this answer; it felt patronizing. I turned back to the sunset and the peace of the beach, wondering if it was time for this pea brain to admit that he had bitten off enough for now. Vlad left me to my silence and gradually, I let the dark thoughts drift away. About a half-hour after the sun had gone beneath the horizon, the sky turned a darker shade of gray and the beach again turned chilly. I suggested to Vlad that it was time to take me back.

CLICK

The remote control button was silently pushed again, and we were back within the white enameled walls of Vlad's Van Nuys home.

"Do you remember?" he asked.

"Yes."

"Everything?"

I answered honestly, "As much as humanly possible." I was now asking myself how I could determine whether this moment was real or a simulation. What test could I use to verify reality? Could I go outside and feel the dirt, rub it between my fingers, smell it and say this is real? The experience I'd just had was disturbing, frightening and beautiful. The thought that there might be a next time bothered me. I could feel and taste small pieces of chive from the cheese. I took another piece and poured more wine. But Comsky had another trick to show me. "Donald, I think there is something you should see now."

CLICK

We were standing in a darkened room with our backs to a wall. As my eyes adjusted, I could make out Ann's bedroom. I watched events unfold in mute horror.

CHAPTER SIXTEEN

Ann had just gone to sleep. The cat was purring at her side. We saw the deadbolt on her door move to the left, and a shadowy figure entered the darkened flat. He found the bedroom and moved towards Ann. "We have to do something!" I said urgently. "We can't let him hurt her!"

"Wait," Vlad whispered. Startled by a sound, Ann opened her eyes to see the dark figure looming above her. Before she could scream, her cat, claws outstretched, was flying through the air. It gave a high-pitched screech as it flew towards the intruder. It sunk its fore claws into his face, hind claws ripping at his throat, canines biting his nose and lips, all at the same time. The man fell back, confused. The cat attacked again, with all the killing instinct and ferocity of its forebears. It was as if it had been bred for this moment.

Mercifully, the cat left the burglar's eyes intact. As he bolted, bleeding profusely, the cat gracefully disengaged from the bloody face and dropped to the floor. The injured man slammed the front door as he fled, lest the beast pursue and attack him again.

Ann lay in bed, stunned. Her pet licked its paws and face clean for a good ten minutes, then slowly, carefully and quietly hopped onto the bed, tentatively, as if asking for permission to be invited back. Ann was still trembling. The big, blind, skinny cat sat at her feet, waiting. Finally Ann said, "Come to me, my love," and the cat lay by her side, purring. Ann placed her hand on the cat's belly, and the cat wrapped her paws around Ann's arm gently and licked it until Ann fell asleep.

I breathed a sigh of relief at the happy ending. Trouble was, I didn't know if we had just watched something happening in the present, the past or the future. "Did what I just saw really happen?" I asked.

Comsky said merely, "Do you think the cat was too hard on the burglar?"

"Are you serious? I think he got less than he deserved. Just imagine attacking a defenseless woman just recovering from brain surgery. Vlad, I will not soon forget that you have the ability to snap my reality around. You showed me a near miss with Ann, and I am both grateful and afraid of your power."

"I thought you would like to know about a significant event in her life, and see it while it was happening."

"So it was in real time? Wow! With you, knowing takes on a whole new dimension. I think I've had enough excitement for tonight."

"One more thing," Vlad said in a matter-of-fact tone. "The man's name is Billy Strakel. He's a sad case. His father was an alcoholic. His mother was a prostitute and heroin addict. He spent his early years witnessing his father beat his mother in a nightly drunken ritual. At least once a week his father turned his rage on Billy. Strakel took to the streets when he was 14. Soon, his body filled out to the stocky shape of his father, though he retained his mother's aquiline nose. He acquired his burglary skills in jail."

"You're not saying we should feel sorry for him!" I was indignant.

"No," Vlad said mildly. "Just understand that when people do bad things, you can guess that bad things may have been done to them."

"We know who he is, let's call the police!"

"Right now Strakel is out of commission," Vlad said. "There's no hurry."

As soon as I got home I went next door to look in on Ann. She was sleeping peacefully with her cat. I touched her warm forehead and stroked the hair away from her eyes, then scratched the cat behind the ear. I went to my flat and prepared for bed, thinking that I might induce a lucid dream myself, but I was afraid of what might come up. I now realized that the time spent on the beach, off the planet and in Ann's home were not real hours, but dream hours. Yet the beach existed, Vlad's planet existed, Ann's trauma and the cat's bestiality existed. After an evening's experience in that "dream state," I was mentally exhausted. If I were back in reality and not in a simulation, and if I believed that Vlad had taken me back to reality, then my question about time in the simulation had been somewhat answered. For the first time in recent memory I went to sleep saying, I do not want to dream and if I dream I do not want to remember. Tonight things were way too confusing. I wondered what would happen the next day when I met with the doctors and Comsky. The last thought I remember having before going to sleep was that I was in a secret office relationship with an extraterrestrial.

CHAPTER SEVENTEEN

The next morning I woke to the smell of fresh coffee, toast and fried eggs. Ann was cooking in the adjacent duplex. Moments later she knocked on my front door. I opened it and she greeted me with a big hug and a cheery smile. She took my cheeks in the palm of her hands, pulled my face down and kissed me on the cheek, and then she looked up at me for a few seconds with her pale blue eyes and seemed to remember something. She smiled warmly. "Come have some breakfast with me, Donald," she said. How could I resist? I followed her into her kitchen. Since her recovery, the rooms had regained the smell of roasted garlic, baked bread and fresh tomato sauce. It was great to have her back. As I sat there watching the miracle of Ann moving around, I had to resist the tears of happiness. Before the Parkinson's, she would tell wonderful stories, make her mouthwatering food and say, "You are such a good eater."

Over breakfast, she told me what I already knew, that her home had been broken into, a strange man had entered her bedroom, and her blind cat had chased him away. She left out the bloody details.

"Are you all right now?" I asked, half-hoping that she thought that it had been a bad dream.

"I'll sleep even better now that I know my cat protects me," she said. As I was leaving, she handed me a small brown paper bag.

"What's this?"

"Just a little something in case you get hungry." She seemed happy just to be able to speak without stuttering. I opened the bag and inspected the sandwich. It was grilled provolone on toasted sourdough. My mouth watered even though I had just finished breakfast. I knew I could use the sandwich to bring me at least mentally back to Ann and my real life in Glendale, in case the situation with Comsky got mentally out of hand. I went back to my home, took a bite of the still-warm, crusty sandwich, letting the rich cheese melt in my mouth. I called work and told them I wouldn't be in today. Then I called my dear friend and confidante Rita Goldblatt, who had first introduced me to lucid dreaming. She agreed to meet me at Peet's in West Hollywood. It was a clear, sunny Los Angeles day. I drove with the window open and sniffed the air deeply. No radio, just freeway and street noises and the real sound of the wind.

CHAPTER EIGHTEEN

Rita was waiting at an outside table in the bright morning sun. Always thoughtful, she had ordered cappuccinos with extra foam for us both. She wore a cream-colored tailored pants suit and a dark cherry red silk blouse, with matching Prada shoes and lipstick. She looked gorgeous, like a fashion model. On her ring finger she sported a simple gold wedding band and a spectacular four-carat, brilliant blue diamond, her engagement ring. Her hair was dyed bright blonde and cut short to frame her delicate features. She reminded me of Eva Marie Saint in *On the Waterfront*, except for the pair of businesslike, black glasses that partially hid her beautiful eyes. She was in her late 40s, and her body was thin and fit. I knew her regimen. Every Monday, Wednesday and Friday morning at seven she did Cross-Fit exercises. On Tuesdays and Thursday mornings it was martial arts training. She was a second-degree black belt.

I shook her soft hand. I knew that her husband was little in evidence; his business interests took him out of the country nine months of the year. Once, after I had become a practitioner in lucid dreaming, Rita noticed a change in my demeanor towards her. I acknowledged the truthfulness and honesty of her observation and my own transparency. I admitted to Rita that during one of my lucid dreaming experiences I had fantasized about her. She did not ask me for details, but suggested that for my own mental well-being, such fantasies were a bad idea for our friendship. I took her sound advice and did not visit her in my dreams again.

"What's up?" Rita asked, adjusting the glasses on the delicate bridge of her nose.

Seeing her, I shook myself from my what-it-could-have-been reverie and got right to the point. "I'm having trouble distinguishing between a dream and reality. Is it possible for someone to actively come into your lucid dream?"

Rita took a sip of her coffee and tapped the table with one cherry-colored fingernail. "There are contests where lucid dreamers try to enter into each other's dreams," she said. "To confirm this actually happens, they are given a word or an object to convey in another's lucid dream. If, when the lucid dreamer wakes up and says that she has obtained the word or can name the object, then it is viewed as a success. But this is still in the experimental part of lucid dreaming and doesn't happen very often."

"So the idea that another person actually enters into the dream, and the entire dream is vividly recalled by each person, is out of the question?"

Rita looked at me a little strangely and waited to see if I was going to say more before she replied. "I have never heard of such a thing happening. However, it is not unusual to invite people into your dreams and interact with them. This has been done many times. You can choose to meet whomever you like in your lucid dream and speak to them as you see fit. The possibilities are endless."

She looked so attractive saying that. My mind momentarily wandered to my fantasy before righting itself. "I'm thinking more as if two people were watching the same movie, the movie being a metaphor for the dream, and then walking out of the theater together, a metaphor for waking up out of the dream, and then discussing the movie. In addition, one of the two people watching the movie had actually made the movie. What do you think, Rita? Have you ever heard of such a thing?"

"I've heard stories, but they usually have to do with ESP. There have been anecdotal stories of similar things for ages."

I leaned in. "So, back to my original question, can you give me a sure way to know that I'm not dreaming and that I am actually in reality?"

"One way that is used by some lucid dreamers comes from a mutual experience that, during the dream state, the dreamer cannot read," Rita said slowly, almost reluctantly. "For some reason, the brain while dreaming has a very difficult time translating the printed word into meaning. As a result, a number of serious lucid dreamers carry a card with the printed inscription: You are not dreaming. Many of them carry it permanently, in their breast pocket. If they find that they can actually read the card, they believe that they are not dreaming."

"Does that work?"

"Seems to," Rita said. "I've used it myself many times. Here, I'll show you." She opened her Gucci purse and pulled out what looked like an ordinary business card. It read, in bold printed letters: YOU ARE NOT DREAMING! "Here, take this one," she said. "I've got plenty." I took the card and breathed a sigh of relief. She looked at me intently, knowing something was going on but she was too tactful to ask. She sipped her cappuccino. "How is Ann doing?"

"She's just amazing. Today, she cooked me breakfast."

Rita nodded. "And how are you doing, Donald?"

I planted a smile on my face. "Just great. I could not be happier now that Ann is back on her feet, and her old self again. I think she is very happy to be able to take care of herself again." Rita looked at me critically. She knew I was avoiding the question. Quickly, I shifted the subject to her family. "How are Benjamin and the kids?"

She shrugged. "Ben is still putting in long hours at the firm, and David and Miriam are now in high school. How is that old, blind cat of Ann's? Is it dead yet?" Rita disliked anything ugly, especially scrawny cats, so I was glad to be able to tell her that appearances to the contrary, that cat was a hero. I gave her Ann's version of last night's event.

"Who would have thought? Is she all right?"

"Just fine, but I think I'll upgrade the locks on her doors."

"You're a good friend, Donald." She put her hand lightly on mine. "You know you can call me anytime." Despite myself, despite my moral convictions, and despite Benjamin and the kids, I felt a little thrill, and Rita knew it.

Rita's card made me feel foolishly invulnerable, as if I had garlic, a crucifix, a bulletproof vest, and had just taken a vaccine against a nasty bacterium. I placed the card in my shirt pocket, patted it a couple of times just to be sure it was still there, and then drove to Venice Beach. When I got there, I took my shoes and socks off, rolled up my pants and walked slowly over the warm, dry sand until my feet touched the cold, wet sand near the surf. I stood for a long while listening to the sounds of the waves, the children playing on the beach, the seagulls squawking. I took the card out of my pocket and read it aloud: I AM NOT DREAMING! A couple of teenagers, passing by, looked at me quizzically. I walked back along the sidewalk, reading the signs on the shops and the menus in the restaurants. Surely, I said to myself, I am not dreaming. I then tried to find the spot where Vlad and I had sat in beach chairs. There were fewer clouds in the sky now. I wondered if the Venice sky had been cloudy the day before.

CHAPTER NINETEEN

At work the next day I discussed another DARPA project with Comsky and the USC docs. This time I was determined to listen carefully and speak very little.

Richard Hausman presented a proposal for the university to participate in a study to determine if a prosthetic arm system could be made commercially available. This meant another cooperative project with the Department of Defense and perhaps the Department of Veterans Affairs, and more money for the university. Hausman was a big man, well over six feet, with the body of a defensive linebacker for Wisconsin, where he'd played in college. In the decade since, he'd added 40 pounds to his middle. Every morning he scanned the Times to compare Madison weather to Los Angeles. In contrast to his size, his voice was almost a whisper. His manner was humble and self-effacing but he got directly to the point. "Apparently, the Department of Veterans Affairs is making available four amputees for this study of our prosthetics program. The Johns Hopkins University Applied Physics Laboratory developed the prosthetic. These sophisticated devices can operate in three-dimensional space. They can reach and grab. Our job would be to develop an interface with the brain's motor cortex."

"What's your plan to do that?" Pebbles asked.

Hausman cleared his throat. "You know we've been developing microelectrode arrays to monitor the synaptic electrical pulses from inside the brain. These arrays have dozens of very small wires that can record synaptic electrical and chemical activity. We will remove a small area of the skull bone and expose the top of the brain where the motor cortex is closest to the surface. We can slowly—that is, very, very slowly—push the array into the brain. If we do this correctly and slowly enough, the neurons will be pushed to the side by the pins. Here is one of the microelectrode arrays I have been considering." He opened a wooden box that was lined with soft foam. Inside was a rectangular chip that looked like an integrated circuit with a tiny dog hairbrush underneath. He passed the box around to the group.

Pebbles pushed his wedding ring up and down his finger. "If we could get this to work, it would be a great step forward, I expect that this brain-prosthetic interface, in addition to working on arms, could also be used for legs?"

"I don't see why not." Hausman said.

Pebbles leaned forward. "Are we equipped to do the computer interface between the microelectrode array and the supplied prosthetic arm?"

"I believe we are," Hausman said. "Of course, there's a bit of luck required when you position the array in the motor cortex. That's the dicey part. Can we count on Dr. Comsky to help with that positioning?"

Vlad, whom I had barely looked at for fear the remarkable events of the day before would be obvious to everyone, replied, "It sounds like this will improve the quality of life for service-disabled veterans and ultimately give them the option of returning to duty. I would have to do the insertion either back east, or have my robotic equipment shipped out here. In any case, I would be happy to assist in this project. It seems also to have a wide application for civilians who have lost their arms. I'm sure their quality of life would also be much improved."

I thought about the veterans' quality of life and how much their lives would have improved if their limbs hadn't been blown off in the first place. I found myself with pen in hand doodling stick figures with lightning bolts running from their heads to their hands. I thought of DARPA's mission to prevent technological surprises, and the motivation to improve the lives of these wounded veterans. There was no doubt that the Veterans Administration and the Army wanted to make their lives better, and also no doubt that they wanted to make America safer and less vulnerable to shock and awe. I put a smiley face on my stick figure.

Pebbles' ring stopped moving. "It seems that if the military can control a sophisticated prosthetic limb by directly taking inputs from the motor cortex, it can use such a microelectrode array to control other physical devices directly from the brain, as well."

"Such as?" Vlad asked.

Hausman responded, "A jet plane, a torpedo, a smart bomb? Take your pick. This technology does have military value."

"I see. We make the brain-prosthetic connection work, and what the military does with the technology is not our concern?" Vlad asked almost in a whisper.

There was a murmur of assent from around the table. Pebbles broke in, "We know that the scalpel may be used for both good and evil, just as nuclear power can be used to destroy or produce clean energy. Should we now discuss the philosophical issues of progress, and bring up everything that could go wrong as we develop new technologies to help the lives of human beings?" He spoke as if he was putting an end to the issue. Looking around the table with a steely eye, he continued, “Gentleman, should we go forward with Hausman's project?”

“We have the technology, resources and ability, and we can do this,” Hausman pressed, his voice surprisingly strong. “We should do this for the good it will do for humanity.” Everyone nodded, including Vlad and me. All were in agreement.

CHAPTER TWENTY

"Want to join me and Mike on my rounds today?" Vlad said as we left the meeting together.

"Where to this time?"

"Compton."

"You sure know how to pick 'em," I said. Couldn't you do your rounds in Hollywood?" I was only half kidding.

"Perhaps tomorrow," he said cheerfully.

The Chevy Impala look grimier than ever. Mike was waiting outside the car with Vlad's satchel. He bowed slightly and opened the front door for Vlad. I got in the back seat. Mike seemed to know exactly where he was driving; I think he trusted his nose to lead us to the most disgusting smells imaginable. When we exited the Chevy, the odor of a four-day-old dead cat permeated the air, at least that's what I imagined it to be. Neither Vlad nor Mike seemed to be affected. I once heard of a man who was born with no sense of smell. He made a career of taking in abandoned or injured skunks. He kept many of them as pets, and none had their scent glands removed. He was happy and the skunks were happy. Holding my nose against the stench, I wondered if skunks are bothered by their own scent.

Maybe, through evolution, their brains interpret their smell as pleasant, but they must know how others react to it.

This time, Mike stayed with us and carried the kit. Vlad did his usual doctoring thing. Desperate people abounded. After a few hours the sky turned to dusk and the air became cooler. There were new disgusting smells—rancid garbage and moldy dead fish—invading my senses, but they weren't as sharp or offensive. My nose was becoming desensitized.

From out of nowhere, three well-built, twenty-something men stepped in front of us. I don't know where they came from, an alley or out of a parked car, but they blocked our way. The smallest of the three asked what we white boys were doing there. They were dressed almost alike, each in blue jeans, black sneakers, and black T-shirts and Dodgers baseball jackets. The speaker had a red bandanna, folded to create a two-inch wide band around his neck. The other two had red bandannas sticking out of their pockets. Was this a gang costume? I took a nervous step back, but neither Vlad nor Mike seemed surprised to see them or in any way alarmed. Vlad was cleaning and dressing a cut on a drug user's hand. I figured he was a user because there was a hypodermic needle lying by his side with fresh blood on it.

The small man pointed to the addict. "Forget about him," he said. "I'm talking to you, old man. You have good stuff in that bag? You white gentlemen have money?"

Vlad spoke calmly. "I'm a doctor trying to help these people."

"These people? These are my people, my customers, and I'm the one helping them."

Vlad looked him straight in the eye. "Are you helping them stay alive by not allowing their wounds to become infected? When they die, you lose a customer. I am also exchanging their needles so they do not infect each other and you lose another customer. You see, I'm trying to keep your customers alive."

"So you're here to help me and my business?" The smaller man smiled ironically, exchanging a look with his companions. "What do you think about paying a toll for the privilege of coming into my territory and helping my customers?"

I glanced at Mike and saw a murderous darkness cover his face for just a moment. "How much?" he asked in a low growl.

"The big man speaks! Well Igor, how does $100 sound?" The thug paused for a couple of seconds, and then added, "Each."

"Each what?" Vlad asked softly.

"You know, I could ask for each of your fingers and toes, or the hairs on your heads, but it's a nice evening and I'm in a good mood. So how about for each of us, me and my two brothers?"

"I can do that," Vlad said, nodding to Mike. "By the way, would you mind telling me your name, so I'll know who gets the toll money next time, because I plan to return."

"Sure Doc, the name's Marvin Gaye, and these are my brothers Hank and William." All three smirked.

Mike reached into his rear pants pocket, pulled out three $100 bills, and handed them to "Marvin," who divvied them up. "Good night, gentlemen," he said. "Keep in touch." Vlad turned back to his patient and I started to say something to Mike, and when I looked up the trio had disappeared. I was trembling. I reached into my shirt pocket and read Rita's card: YOU ARE NOT DREAMING. Vlad finished bandaging and said to the addict, who sniveled his thanks. "You're going to be fine," Vlad said. I felt he was talking to me.

CHAPTER TWENTY-ONE

Before Mike dropped me off at my car, Vlad asked if I wanted to go to his place for a drink. I accepted eagerly. I had many questions to ask. In his immaculate kitchen Vlad opened a bottle of chilled Rombauer Chardonnay and poured some of its oaky contents into two Riddell wine glasses. He placed a tray with the wine, a wedge of soft Brie, and crisp English crackers topped with sesame seeds on a small table that was between the two chairs facing his large flat screen TV. He held up his glass. "To life, which is good." I instinctively raised mine, touched his glass, liking the clear bell tone. The delicate smoothness of the wine relaxed me. It brought back pleasant memories of summer evenings in California wine country. I'd learned to appreciate good vintages on tasting tours throughout the Napa Valley. Where had Vlad learned his wines? Were there vineyards on Paucri?

After a few minutes, Vlad said, "I'm sorry you were frightened this evening. You were in no danger."

"How do you figure? Those men could have had guns."

"Actually, Donald, they did have guns."

"So, Vlad, what's not to be afraid of?" I asked.

"It's complicated."

"I know, and I'm incapable of understanding," I said sarcastically.

"Do you want to hear a story?"

Too nettled to answer right away, I cut a small slab of the Brie, placed it on a cracker, took a sip of wine, ate the cheese and cracker, and took another sip of wine. "Sure, I like stories," I said. "Am I supposed to listen, and not ask any questions until you're finished?"

"I'm here for you, and you can ask me whatever you want, whenever you want."

Mollified, I waited.

"What I want you to understand, Donald," he began, "is that the human mind yearns to be elsewhere in a different reality, even without the stimulation of sound or video. Consider the novel: the intimacy of the words and the magic that they create in the brain from symbols on a page, bringing to life characters and their stories. Emotions are felt and the reality is created in the brain. For gifted writers, the ability to conjure up imagery and sensory perception with the printed word is an astonishing achievement.

"Consider the ancient Greek plays, where the audience is lulled into believing in the characters being portrayed as they mouth the words of the playwright. The audience is swept away from the cares of their world into the world of the playwright, through the actors on the stage. Magic happens. The works of Homer and the poetry of Beowulf took people to a different world, a world of imagination and epic tales.

"It is part of the basic nature of the human mind to crave to leave reality and substitute their day-to-day experiences with the world of imagination and the lives of others. Through television, the Internet, sportscasts, newscasts, the political realm, the involvement in other people's lives, reality TV, nature shows, documentaries, science programs, animated voyages to the moon, Mars, and the planets of the solar system, we are transported elsewhere. The list is endless." Vlad paused, as though considering whether to continue. "But why tell you a story if I can show you in a simulation?'

"Even better," I quickly answered, taking a gulp of wine.

CLICK

We were alone in our lounge chairs at Venice Beach an hour before sunset. The scene was different than before, there were fewer clouds and the air was warmer. I sat there quietly for a few minutes, then reached into my shirt pocket and pulled out my card and read it. It did not lie. I was not dreaming; an alien had taken over my entire conscious perception of reality, with my permission. I was not afraid. I would let him do it again. "Okay Vlad," I said. "I want to see things from your perspective. I like this beach scene. It's terrific."

"How do you like the wine?" Vlad asked suddenly.

I regarded the glass in my hand. "It's as good as it ever was. Why?'

"You know this is a simulation, and you are drinking simulation wine. You can drink all the wine you want here and you will never become intoxicated. You can eat all the food you desire and you will never put on weight. All the food has zero calories because it is not food."

What an idea for a diet! Ever since I was a kid, I'd struggled to keep from putting on the pounds. I loved food too much, that was the trouble! "I could get used to this," I said enthusiastically.

"But it gets better," Vlad said. "Take another taste of your wine."

I held my glass to the setting sun. The wine looked the same, a very pale yellow, but when I sniffed it, it reminded me of dark cherries. I took a sip. Zinfandel. "You did that to the chardonnay?" I squeaked.

Vlad looked pleased with himself. " I could make that wine taste like anything I like, or more importantly, anything you like."

I took another sip. The chardonnay had returned.

"Are you ready for some more excitement?"

"Sure," I said, although I wasn't at all sure.

"I am going to show you this instead of just telling you about it. You might want to put your glass down."

CLICK

I was glad I did as he suggested because Venice Beach at sunset disappeared, replaced by Pebbles' office this morning, and I was there with everyone else from the meeting, with a major difference. We were frozen, looking at ourselves as if we were wax figures, including Vlad. My right hand started to reach for my shirt pocket and the card. The thought evaporated. "Wow," I said, and the sound came out like a 78 record at 33 speed. Vlad stepped in front of me, as if to break the hypnotic hold that the scene had produced.

"What exactly am I seeing here?" I asked.

"Look around, walk around, touch whatever you like. You won't be disturbing anything."

I felt as if I were in the opening scenes of 2001 A Space Odyssey. I was the chimpanzee who had just discovered the large black monolith. Like that chimpanzee, I walked around the room slowly, carefully and quietly. I approached my seated self, reached out a finger and touched my nose. To my surprise it did not feel like skin; it had no give; it was hard as stone. I tried to pick up papers from the table, but they would not move. "Yes, Vlad, I can see," I said. "I won't be disturbing anything."

"That's right. This is the past, and in this simulation, you cannot change anything. This is a computer-generated simulation based on the visual experience of all the participants you see here, frozen in time. Would you like to see a different view?" I nodded.

CLICK

I was in my body at the meeting, looking across the table at Pebbles. I was paralyzed. I could not move my eyes or any part of my body, and of course, I could not speak. I felt as if my arms and legs were in stocks. I was claustrophobic in my own frozen body and I was not happy.

Pebbles was speaking. "It seems that if the military can control a sophisticated prosthetic limb through the use of directly taking inputs from the motor cortex, it can use such a micro-electrode array to control other physical devices directly from the brain." I found myself doodling on a paper before me. I tried to put the pencil down, but I had no control over my hand. I seemed to be locked into myself, seeing and hearing or, I should say, re-hearing and re-experiencing the morning's meeting. I drew a happy face on my stick figure. I watched my hand move. The happy face had no relation to the claustrophobia I was feeling.

CLICK

I rubbed my eyes and looked around the room. My body felt lighter. I was standing next to Vlad again. The room was still full of wax figures, including mine. I knew if I touched them they would feel like stone. I shivered. I remembered times in my life being so deeply asleep that after I woke I couldn't move my body. I had tried to move, to talk, and for long moments felt trapped and helpless. That's what this was like, and I hated it.

CHAPTER TWENTY-TWO

I became aware that Vlad was looking anxiously at me. "Are you all right?' he asked.

"That was . . ." I struggled to find the right adjective, "disorienting. I prefer this view."

He smiled an enigmatic smile. "Donald, I wouldn't put you through these things if I didn't think you could handle it. I wanted to show you the source of the simulation you just experienced. The sensory perceptions, in the form of electrical signals from the synapses, are recorded and then interpreted by a computer. That is the way the past is recorded. In this computer simulation, we are free to observe and walk around in what you may think of as a hologram. Let's go stand over here behind Pebbles."

CLICK

All the people in the room became animated. Pebbles was talking, I was thinking about being a fly on the wall.

CLICK

We were back at the beach having crackers and wine, but this time I wasn't looking at the sunset, or the clouds, or the sandpipers, only at Vlad. "You really are an alien," I said, raising my glass. "To life, which is good." We drank solemnly.

I watched the sandpipers darting in and out of the surf line. The whole scene was hypnotizing. It might be possible to stay there forever and not use much real time, I realized. I believed I knew what I was dealing with now: I could have a relaxing life at the beach without losing more than, what? A minute out of my life? The wine was delicious, but completely non-intoxicating, and I wasn't sure how I felt about that. Then I wanted to kick myself for thinking about the effects of wine when I had just experienced traveling to the past. Time travel! This is what Vlad meant when he wanted me to see from his perspective. He had access to a recorded past because of his alien technology. I imagined that he took his power to travel through time for granted; he'd done it so much. I imagined him as that fly on the wall, having the answers to questions like: did Lee Harvey Oswald act alone? Did the Presidential Book of Secrets really exist? I tried to think of a question I could ask him that wouldn't sound stupid, but then, every question I asked would be stupid, because he seemed to have all the answers at the tip of his tongue.

"Vlad," I said, "I'm trying to get my head around what's going on here, and because I don't understand your alien technology and probably never will, let me tell you what I think I understand. Okay?"

Leaning forward on one elbow, Vlad offered me a cracker with cheese, which I declined. "I'm listening," he said.

"I'm thinking about the black monolith in 2001. The image sticks in my mind. It represents an unknown and mysterious alien technology. That's what you have brought to me, an unknown and mysterious alien technology, and I choose to look at it like the black monolith, except—and I know this sounds silly—I don't like the monolith's sharp corners. I prefer something like a sphere, a Buckyball."

Vlad nodded. "The geodesic dome."

"Buckminster Fuller's, yes. I see this ball at a microscopic level, and it contains mysterious alien technology. I don't know what it is, and I don't believe I can understand it, but it exists inside this Buckyball, at least in my mind, and I know what it does. It can record the electrical impulses that exist within my synapses, and it can also inject an electrical impulse into my synapses. This is the way your technology has gotten into my brain and allowed me to experience the beach, the cheese and wine and crackers, and now, time travel." I glanced at Vlad to see if I was on the right track. He looked at me as if we were in a classroom and I was his star pupil.

Gaining confidence, I continued. "There is a computer interface between these Buckyballs in my body, creating this dreamlike simulation that seems absolutely real to me. You took me into orbit around Paucri. If you can do that, I bet your simulation machine can take me just about any place I can imagine, and once I'm there, I would have no way of distinguishing that from reality."

"That could happen," Vlad admitted.

"I can see losing myself in this Venice Beach simulation and not wanting to come back to the real world." I picked up a handful of warm sand and let it fall through my fingers. "Have I got it right? Is that how your alien technology works?"

"I like the Buckyballs, they are a great metaphor. And yes, there is nothing significantly wrong in how you imagine my technology. You seem to have accepted it, as a passenger in a car may not understand the details of a fuel injection system. So long as the car gets you where you want to go, you don't need to know how it works. A fuel injection system by any other name: what if we called it a Buckyball system?"

"Why not? But just for the record, what do you call it?'

Vlad gave me a kind of half smile I'd never seen before. His mouth tilted way up on the right side of his face, and his eyes twinkled. "On this planet, in your language, we call that synaptic alien technology interface unit a Buckyball."

I laughed. People thought it would be hard to communicate with aliens, that we would have to start with something really simple, say, a numbering system, but here he was speaking English as well as anyone, fitting right in, and making his technology understandable even to my pea brain. I was sitting at Venice Beach in a dream world with an alien, and I was at peace.

CHAPTER TWENTY-THREE

With Vlad's help, the new grant application for DARPA was completed ahead of schedule and was acceptable to Dr. Pebbles. Ann was as happy as I had seen her in years. With her guard cat, she was in good hands, or more to the point, good paws. It had been a stressful number of weeks but things were peaceful now, and I felt I needed a vacation. Mentally, I was already there. I approached my travel consultant.

"Vlad, you asked me where I would like to be and I chose here at Venice Beach an hour before sunset. Now let me ask you, what's your favorite place?"

"I've got many favorite places. You have been to one of them."

"Pebbles' office?"

"That's good, Donald. You are making jokes with an alien."

"Paucri?" I suggested.

"I thought someplace closer to home. Anaheim?"

"Disneyland?" I almost asked how he knew I'd been there and loved it, but stopped myself, because of course he knew.

"Want to go?" Vlad said, sounding like he was asking a dog if it wanted to go for a walk.

"Who wouldn't want to go to the happiest place on earth?" I answered. If I'd had a tail, I would have wagged it.

"Let's do it for real," Vlad said. This confused me.

"Isn't it a little late? I mean in real time? What time is it in real time, anyway? Does my watch display the real time and date?" I looked at my quartz watch. It read 11:55. Here at Venice Beach, it was just pass sunset. With all that had happened in this simulation, I was unsure what time meant, if it was close to noon or midnight in the real world.

Vlad said, "It's close to midnight. You have had a busy night and you have been drinking. You can stay here with me. I have a guest room."

"You're trying to confound me. You don't really mean here, in this simulation, do you?"

Vlad looked contrite. "Forgive me; I was not trying to confound you. I think it is great that you understand what I mean and do not take me literally. So, how about it?"

"How can I refuse? Yes."

I awakened the next morning to the smell of rich coffee. Vlad, in his kitchen, was preparing an omelet and toasting slices of whole wheat bread. His kitchen, like the rest of his house, was painted in white enamel. The fixtures, sink, stove and refrigerator were all brushed stainless steel. A commercial chef would have been happy to cook in Vlad's kitchen.

Over breakfast, we talked about a bionic eye, an artificial silicon retina implant now in use. Not the usual light kind of breakfast conversation, but after all, the last breakfast conversation I'd had was with Ann, who gave me a graphic description of her cat brutalizing Billy Strakel. Vlad served the omelet on a heated plate. It was slightly crusty on the outside but when I cut into it, a molten stream of cheese, grilled red peppers and onions flowed out. The steamy aroma of the cheese, onions and peppers was heady. I took a bite and was not disappointed. The toast was golden brown and full of different seeds. I spread soft, unsalted butter over it and it melted instantly. I bit into it. It was delicious. The coffee was perfect. “Vlad, this is great,” I gushed. “Did they teach you how to cook at alien school?”

“You know, Donald, you should be careful about referring to me as an alien here.”

“Oh I get it. The Rio scale.”

He grew serious. “I am so glad you like my cooking. Donald. Now, may I make another confession to you?”

“As long as you're not going to take me away from my delicious breakfast and put me into another one of your simulations. I'm enjoying this way too much.”

“No, I'm not going to take you away from your breakfast. Believe me; I know how much you are enjoying your breakfast. That is just it. I cannot take full credit for how your breakfast tastes. Well, that's not true either.”

He had me puzzled now. “What are you getting at?” Had he or had he not cooked the breakfast? I wasn’t in much of a hurry to get an answer. I piled some omelet onto toast and slowly chewed it, savoring the flavors.

“You tasting and smelling this breakfast is a simulation,” he said rather regretfully.

I stopped chewing in mid-bite. “This is not real food?”

“The food is real enough and I did cook it. However, what you are tasting and smelling is a simulation,” he repeated. “Want me to turn it off?”

I tried to relish how good everything tasted. I said, “Hold on, let me finish this mouthful and enjoy this coffee.” I hoped I could tell the difference between real and simulated. “Okay, Vlad,” I said resignedly. “Do your magic.”

CLICK

The half-eaten omelet lay before me. I took a forkful and bit into a piece of toast. Everything smelled very good but the omelet was no longer the best omelet I could imagine. The same was true of the coffee, which was full-bodied but not great. "You're still a very good cook, but not a superb one," I said a trifle peevishly. "Can you please take me back to the mouth-watering simulation?"

CLICK

CHAPTER TWENTY-FOUR

Vlad was right. I had been to Disneyland many times over the years. We arrived at 11 o'clock in the morning. He was wearing the same working clothes, a beige shirt, pants and boots that he'd worn at our first meeting. They brought back memories that seemed distant, perhaps because so much had happened since that night in the basement computer room. I thought how natural it was for Vlad to choose Disneyland, which was the Hollywood interpretation of a simulated environment: human-sized mice, dogs, princesses, Mad Hatters, flying elephants and magic castles created for the visitors. How easily humans accept this artificial environment!

We presented our tickets at the gate and strolled in. "Why is this your favorite place?" I asked.

"It's such a great place to meet people from all over the world," Vlad replied. "The children are joyful and safe. It's a place where their dreams come true. Do you want to see what I do here that makes me so happy?"

"Of course, that's why we're here."

"Follow me," Vlad said.

We took the Disneyland train to Fantasyland. He bought a red balloon and we wandered down a pretty path and sat on a bench. After a short time a young girl, perhaps seven years old, walked up to Vlad and pointed at his balloon. She was wearing a white dress, black leggings and red sparkly shoes. Her hair was brown and it had corkscrew curls in it, like Shirley Temple's. Vlad handed her the balloon and began speaking to her. Although he spoke slowly and clearly, I couldn't understand a word he was saying. Her face became animated and she smiled as she answered Vlad. This time I understood one word: Greta. He called her by name and asked her a question that I couldn't make out. Greta pointed behind her and shrugged her shoulders. Vlad pointed to me and said slowly, as if speaking about an idiot, "Donald." Greta nodded at me and pointed again over her shoulder. The two continued to speak. Greta looked sadly at me, and I began to feel like an idiot. Nothing they said made sense. Greta, with balloon in hand, sat down. Companionably, we watched the park-goers parade in front of us—children pulling on their parents' arms, parents struggling to read park maps, costumed blond princesses surrounded by chattering children, parents snapping pictures of laughing children with various Disney characters.

"Greta!" A young, blonde woman broke through the crowd, came to a stop in front of Greta, who jumped down from her seat and hugged her. The woman was dressed in jeans, a yellow shirt, and red tennis shoes, and her eyes were brimming with tears. She stood hugging the child, who had wrapped her arms around her neck while keeping a tight grip on her balloon string. The woman called out, "Nikki, Nikki!" and a young man emerged from the crowd, hurrying forward. Both parents embraced the child and each other. There was animated conversation and finger pointing by all three. In the end, Greta pointed towards me. The man and woman, holding their daughter's little hands, approached. Vlad stood up. The four of them spoke for a few minutes before Greta again pointed to me, saying,

"Donald." The couple looked at me sadly. Nikki reached out his hand. I stood and shook it.

"T'ank you, t'ank you," Nikki said, putting his arm around the woman's waist. "Wife Teresa." Teresa beamed at me and we shook hands. She thanked me too, and we all smiled. They bid me a special goodbye. Teresa warmly hugged Vlad, and then she hugged me. I was not exactly sure what had happened but I recognized a happy ending. I sat down and watched the red balloon disappearing into the crowd. "So this is what you do?' I said.

"It is."

"It's all very nice, but I feel like a bump on a log. Are you going to be saving lost girls all day? And what was that language you were speaking?"

"They're from Switzerland and speak a dialect of their canton. I thought they would appreciate speaking to someone who could make them feel at home."

"Let me guess," I said. "You are using your simulation to understand and speak a foreign language? Something like you did this morning at breakfast with the omelet?"

"Mr. Thomas, you catch on fast."

"Well, are you going to let me participate, as you did at breakfast this morning?"

"I thought you'd never ask," Vlad grinned.

CLICK

At first, nothing seemed to happen. I sat there with Vlad watching the crowd go by, listening to snippets of conversations.

After a while Vlad asked, "Are you hungry? Want some ice cream?"

"Sounds good, wait here while I get some. There's a cart right over there."

"Make mine chocolate," Vlad called.

The ice cream vendor, looking very bored, dressed in blue and white vertical-striped pants, matching hat and white shirt, stood beside his freezer cart. I said, "May I have one vanilla and one chocolate ice cream, please?" The young man straightened as if surprised, and shot me an enormous smile.

"It's good to hear you, my friend. How did you know I was from Croatia?"

I stared, puzzled. He continued, "Sometimes I forget where I am and it's difficult for me to speak in English. How long are you here for? Who are you with? Are you alone? It is so good to hear my mother tongue again."

I wasn't sure how long I could keep up the act. I wanted to get my ice cream and return to my bench. Oh that Vlad, what a sense of humor! "I'm just here for the day," I said in Croatian, and pointed over towards Vlad. "I'm here with my friend, who's hungry for his chocolate ice cream. It was very nice to see you, my friend."

"You don't know how you made me feel homesick," the man said, and waved me away when I tried to pay, saying, "It's on me, my brother."

Vlad smiled as I handed him his ice cream and sat down.

"You are good," I said. "And yes, I'm beginning to see more from your perspective." We sat there eating our ice cream while I intensely listened to people passing by. All were speaking English. After a while, I focused intently on their lips and noticed that just as in a dubbed movie, their lips didn't match the words. When they did match, I believed they were actually speaking English. This new sensation of being in a dubbed movie was intriguing, but eventually, my thoughts drifted towards the Rio scale. If Vlad had been doing this for some time, people would believe him to be a gifted linguist rather an alien. But how about me? Vlad had me speaking and understanding other languages. How could I explain this to my friends? How could I explain Vlad to my friends?

CHAPTER TWENTY-FIVE

Vlad had long since finished his ice cream treat; he was holding a crushed wrapper and wooden stick in his lap. "Ready for a walk?" he asked.

We left the quiet path and moved into a crowded plaza, dropping our papers into a wastebasket.

"Vlad, I have a few questions for you, but this doesn't seem like a good place to speak. There are so many people who could overhear our conversation."

CLICK

"I've taken care of that, Donald," he said smoothly. "In addition to our current ability to understand everyone around us as if they were speaking English, I have arranged that when people hear us, they will not understand us."

"Really? What language will we be speaking in?"

" Inuit. There are no Inuit visiting today."

"But won't they see that our words and lips don't match?"

"Are you serious, Donald? Look around."

We were in the middle of a large crowd. The droning sound of countless voices surrounded us. The noise rose in a babbling hum. Vlad raised his voice. "All you people look at me! I'm an alien from another planet!" He opened his palms to the sky, as if to say, "why aren't they taking me away?" No one looked twice at us.

"I get your point," I said, "but it's not a fair test. With this crowd, standing here in Tomorrowland, you could be speaking in a language everyone could understand and not be taken seriously. Let's keep speaking Inuit, though. It makes me feel relaxed. Unless someone I know happens to hear me."

Vlad shook his head. "Among other things, you don't know anybody here. Want to take a ride on the big house that floats on water?" That apparently was the best translation for riverboat in Inuit. Not waiting for an answer, Vlad turned and headed towards New Orleans.

Considering the thousands of people at Disneyland that day, how could he possibly know that I didn't know anybody? But I believed him. "Do you know everything about me, Vlad?" I asked, perturbed.

"I don't know what you're thinking."

"Well, that's a start. You want to know what I'm thinking?"

"Anything you would like to share with me. I am not a mind reader."

We had almost reached the New Orleans section of Disneyland. I could hear a band playing Dixieland not far off.

"Right now I'm asking myself why this creature from another planet should care about my thoughts. What's so special about me?" I said.

Vlad said, "Do you remember that one of your first questions was about my 'alien technology'? I asked if you had heard of lucid dreaming. I know you have been doing that for years. Your mind has accepted that you not only remember your dreams; you manipulate your dreams. There are few people who can do this, and fewer are involved with the medical research profession, still fewer with DARPA. I sought you out to make contact, and have done my best not to freak you out. How am I doing?"

Even though I had won this honor by default, given there were so few choices to pick from, I was flattered. "You're doing fine on the not-freaking-me-out scale," I said, beaming. But please, why were you so interested in knowing what's on my mind?"

"I'm trying to make contact with an alien species," Vlad said as though this were the most obvious reason for asking another's thoughts. "That's you, Donald Thomas. You are distinct from me. I am trying to communicate with you, and I would like to know how well I am doing. My people have different meanings for the words Earthlings use, so I do not want to assume how you are reacting. So I ask. I'm trying to establish a relationship with someone very different from me."

It struck me that no one knows what anyone else is thinking, and asking was required in all relationships, but I didn't say so. "I'll tell you when I was freaked out. We were in the simulation in Pebbles' office, and you put my consciousness into my frozen body. I could see, but I was paralyzed, and I was seriously upset. It felt a little better when you allowed my frozen body to become reanimated, even though I had no control over it. I was a passenger inside of a robot body that had a mind of its own. You did not leave me in either of those uncomfortable states for very long, but it was enough to give me a taste, and I was unhappy there."

"I think I understand," Vlad said, "and I have experienced what you describe. Although such a state has not bothered me for a long time."

"I expect that seeing things from your point of view will make me uncomfortable at times," I admitted. "I expect that your further confessions to me will make me even more uncomfortable; otherwise, you would not use that word, "confession." Am I correct? How am I my doing, as far as sharing my thoughts? And yes, Vlad, I would like to communicate with you also. So I ask for and need frequent validation." Vlad nodded, and we moved with the crowd toward the riverboat landing, trying not to get in the way of anybody's perfect picture. I got a kick out of watching lips out of sync with voices. I asked myself whether this was a metaphor for what it was to be like in a relationship with an alien.

"Should I continue to tell you how strange all of this is to me, watching people lip-synching English? Or should I just tell you when it's no longer strange to me?"

Vlad didn't answer. "I think it's only fair that you should know my thoughts," he said suddenly. "I expected that you would be uncomfortable in those two situations in Pebbles' office, especially since you could not speak to me. You are correct. I wanted to give you a taste of that experience so that you would know firsthand how the simulation obtains its information. May I use the word 'firsthand' about a simulation?"

I nodded.

"There is no need, then, to go back to those perspectives in the simulation. Now, as to your question, please feel free to tell me when things no longer feel strange to you. And I do have much to confess, but I have to be in the mood, and I'm not in the mood right now. We're in Disneyland and I want to have fun."

CHAPTER TWENTY-SIX

Like a kid, Vlad eagerly claimed seats on the top deck of the riverboat. There was a loud whistle and slowly, we pulled away from the dock. "Can we eliminate this language simulation now?" I asked.

CLICK

The suddenness of it bothered me. "Can you just turn that on and off anytime you feel like it?" I asked, shaken. Everyone's lips now moved with their speech, and though I couldn't know for sure, I was no longer talking Inuit.

"Yes I can," Vlad said serenely.

We were passing Tom Sawyer's Island. I doubted whether Vlad really wanted to hear all of my thoughts, my total stream of my consciousness. A country-western lyric came to my mind, A little less talk and a lot more action. I remembered the invading aliens in War of the Worlds, Independence Day, Starship Troopers—aliens with ray guns taking our minerals, our water, the entire planet, killing everyone who stood in their way. This made for an exciting story, a real page-turner, but here I was with Vladimir Comsky, a quiet surgeon who had some sort of simulation machine that, among other things, let us travel to the past. I had done it, and seen that one cannot change the past, it is literally cast in stone. He said he wanted to have a relationship with me, but how well did I know him? How well could I know him? Would a life with Vlad mean spending my evenings going to the dirty corners of the world helping sick people, exchanging needles with drug addicts? Would a life with Vlad mean endless visits to his simulated world, a life lost in fantasy, perhaps debauchery and gluttony? Would a life with an alien mean exploring the past and obtaining answers to history's mysteries? That part excited me, of course. The prospect of his future confessions was troubling, though. Perhaps he had evil intentions, and those horrible sci-fi movies were not far off the mark. They served their purpose as a warning not to be taken in by the stranger, the one who is different, especially the one who is looking for a relationship.

I wanted to feel good about my friendship with Vlad. I wanted to put aside misgivings and mistrust. But first I needed to calm down. I raised my eyes to the blue sky, where puffy white clouds drifted, and then to the children playing on the island who were surely making new friends, and thought, the world has changed because of this alien, but in a good way. Greta's parents found their little daughter, attracted by the red balloon. Ann was cured of her Parkinson's. What was wrong with me? Was he trying to fix something in me, too? This man, this alien, had been nothing but patient and good with me, and I had shown him mistrust. Impulsively, I blurted, "Vlad, can you do something for me?"

"Ask," he quickly replied.

"I want to find the man who tried to attack Ann."

"Billy Strakel?"

"You know where he is, right?"

"Yep."

"You know where he is right now?"

"Yep."

I saw that I could have a conversation with Vlad all by myself. I'd ask the questions and answer with his standard three-letter words. Annoyed, I said, "Why are you talking like that?"

"Like what?"

" Like yep.'"

"I like the terseness of it. To the point."

"Well, it's silly. Montana cowboys talk like that. We're in Anaheim. They don't say 'yep' in Anaheim."

"All right, Donald. I'll try to remember where I am."

"It would be okay in Montana," I grumbled, a little ashamed. The guy knew all the languages in the world—he was allowed to make a few mistakes!

I returned to the subject of Strakel. "I want to stop him from doing any more bad things."

"You want revenge?"

"Is that wrong?"

"You know he never touched Ann, and his face and head are all scratched up."

"But he was going to, wasn't he? He was going to do bad things, wasn't he?"

The riverboat had docked. We stood. Vlad said, "Yes, I believe he was. Sounds like you have had enough of Disneyland for a while. Want to go back to my place and see what Billy Strakel is up to?"

"Yep," I said. Vlad laughed. When he laughed he opened his mouth wide and threw back his head, showing his teeth. No trace of a filling, all perfect white. I wondered what his physical form on Paucri was, and how he chose the way his human body would look.

After this, I half-expected to be in Vlad's living room the next instant, but we drove through the Los Angeles basin to Van Nuys. I could feel the temperature change with every mile, including the final heat increase in the San Fernando Valley. The first thing Vlad did when we entered his home was to turn on the central air conditioning. The first thing I did was to hop into his reclining chair in the living room. “Let's go,” I said. “Fire this thing up.”

CLICK

CHAPTER TWENTY-SEVEN

We were standing in a small room somewhere. The shades were drawn, and it took a moment to get used to the gloom. There was an unmade bed against the wall, a straight-back chair and a small table near the window. A light shone from the open bathroom and I could hear noises from the sink. I went to the bathroom doorway and saw Billy Strakel at the mirror applying fresh bandages to his face.

Vlad said, "What you want to do now?"

"What are my choices?"

"Just about anything."

"You mean I can kill him?"

"Sure, but then you would be a murderer."

"What do you think we should do with him?"

"Let nature take its course."

"You mean just do nothing?" I heard humming from the bathroom, and then Strakel said, "That damn cat, I'm going to kill that damn cat."

"Did you hear that? He said he's going to go back for revenge!"

"Well, we can't just let that happen, can we, Donald? Did you know that the police have an arrest warrant out for him? He is a suspect in a number of burglaries but they do not know that he is also a serial rapist. If the police arrest him now, they may not have enough evidence to convict. The police have no fingerprints, eye witnesses nor DNA evidence."

"But what about the attack on Ann? All those scratches on his face and neck?"

"He'll say he cut himself shaving, and without any evidence, a public defender will get the case dismissed."

My voice rose. "So what good is all your technology? We know he's a criminal, we know where he is, we know he is going to continue burglarizing and raping. Don't you aliens have any morality? Don't you punish people who do bad things, especially when you're absolutely sure that they've done them and will continue to do them?"

"Now Donald, I want you to see things from my point of view," Vlad said soothingly. "Strakel is not one of my people, he is one of yours. He is your problem, now that you know he is guilty and that he will commit more crimes, do you want to act as judge, jury and executioner? Do you want to take all that responsibility? Do you want to round up the Strakels of the world and kill them all?"

"No, I guess not," I had to admit. I looked into the bathroom. Strakel actually looked pathetic with bandages all over his face and neck.

There was a loud knock on his front door. Strakel turned away from the mirror and looked directly at me. I was caught and now he would attack me, I was certain. Where was Vlad? I took a fearful step backward, and Billy Strakel walked directly through me. Nauseated, I felt my knees buckle. I clutched the doorframe and turned to see Strakel opening the front door, taking money out of his pocket, passing it through the doorway, and returning, holding a condom with a knot tied in it.

"Heroin," Vlad said, behind me.

"Where were you just now?" I demanded.

"Right here," Vlad said. "When you're afraid, Donald, you don't see as well, do you?"

We watched Strakel prepare the heroin, inject himself between his toes, and fall into bed in a stupor. "I've seen enough. Let's get out of here," I said.

CLICK

And just like that, we were back in the cool, white, air-conditioned living room. I was exhausted. My brain was having difficulty remembering which parts of the day were real and which were simulations. And then, there was the blending of the real and the unreal—tasting an enhanced omelet, speaking Inuit and hearing everyone's language as English. And my brain was still trying to process what had happened when Strakel turned from the mirror and walked right through me, though I consciously realized that the entire trip to Strakel's hovel was a simulation and his body passing through mine was an illusion.

"Vlad," I said, "I'm a little freaked out." Understatement. My brain was screaming that it was a nightmare.

"Tell me about it."

"I keep thinking about Strakel walking through my body. I can't get the image out of my head."

"That's a tough one to accept. How did it feel?" He's feeling scientific curiosity, I thought with a shudder.

"I saw it, I experienced it, my body reacted to it, and my brain remembers it. But I didn't feel a thing. Nothing."

Vlad laid a comforting hand on my shoulder. You pet the mouse after you inject it, I thought. "I may be able to help you with this," he said. Do you want me to try?"

"Please."

"We will have to reenter the simulation. Are you okay with that?"

I took a deep breath. "Okay, if we have to."

CLICK

Nothing appeared to change. We were still in Vlad's living room. He stood before a blank wall.

"Come over here," he said. I joined him.

"Feel the wall, the texture, the firmness and the seeming reality of this simulation," he said. "You know I am controlling how this simulation works, don't you?"

I gulped and nodded.

"Put your hand on the wall, in fact, put both hands on the wall, lean into it and notice how it supports you. Now, feel how it turns to soft clay so you can move your hands in it. Now take your hands away and lean against the wall again. It's back to being hard. Now here is the good part. Put your hand on the wall again and move it forward."

I did as Vlad asked, felt the wall becoming clay, then hard, and then it dissolved. My hand went through it like it was nothing. I put my other hand in the wall. I waved both hands around and felt nothing. I turned to Vlad, elated. He took my right wrist with his left hand and asked me to stick out my index finger, moving slowly to face me directly.

"Relax," he said. "This won't hurt."

I relaxed my right arm and extended the index finger. He moved it between our faces, then brought his right hand up, and with an open palm moved my finger slowly through his palm, so that the tip of my index finger protruded from the back of his hand.

"How does that feel, Donald Thomas?"

I wiggled my finger. "I don't feel any sensation. I think I could close my eyes and not know your hand was there." I withdrew my finger and swept my hand over Vlad. There was nothing there. Vlad was an illusion. All of it was an illusion, a dream, and nothing frightening. I felt my body accepting this and the tension drained from me.

Vlad said, "Are you ready for graduate work?"

Without hesitation, I stepped forward and walked right through Vlad's body.

CLICK

CHAPTER TWENTY-EIGHT

I was back in the comfort of the recliner. My brain had undergone one of those ah ha! moments. I was no longer dwelling on the Strakel visit. I was thinking of all the other things that had happened, especially the conversations with Vlad. I was aware that I used to think about other things; I was involved with work projects and with my neighbor Ann. Now I was thinking about time travel and revenge. And I wasn't sure I liked the change.

"I think I need a vacation," I said wearily as Vlad brought in a bottle of wine, a bowl of Kalamata olives, hummus and pita bread. "Even though I have completed my graduate work, my head is still swimming. I think you've got me out of my comfort zone. Up to now my life has been fairly conventional. I've been pretty much a non-activist moving through life comfortably, not really liking change. I think I've been over-stimulated by your simulation machine." I tried to smile. "Would you mind if I just went back to my normal life for a little while and let my brain settle down from this simulated reality of your life?"

Vlad spooned hummus on a pita triangle and handed it to me, then uncorked the wine and poured me a glass. "Of course not, I think it's the right thing to do," he said. "I can see that my life might feel like a non-stop magic show. I will be around, doing my thing with Mike."

Gratefully, I took the glass of wine and sniffed it. "Nice aroma," I said. "Figs and cherries."

"On second thought, Donald," Vlad said, swirling his glass, "because today is Thursday and you deserve a break, relaxation and entertainment, let me treat you to lunch tomorrow at the Academy of Magical Arts in Hollywood. You probably know the Magic Castle over on Franklin Avenue. What do you say? I'll ask Mike to pick you up at 10:30 and I won't use any simulation on you unless you request it. How does that sound?"

I sighed. "Well, that sounds like a change of pace, anyway. It might be nice to see real magic from real humans."

Maybe a good night's sleep would be the only break I needed. Besides, I could always change my mind about Hollywood and tell Mike and Vlad to go without me. In any case, I could sleep on it. When I got home I Googled the Magic Castle and found that not only were a jacket and tie required, but Friday was a members-only luncheon. So not only was Vlad a member, but he had already arranged for me to be his guest. I felt irritated, but it was no wonder he took me for granted. Had I ever refused to go along with him? I just couldn't seem to say no. My mind was exhausted, yet I wanted to go on the roller coaster ride again. I was beginning to think that life with an extraterrestrial, or at least with this extraterrestrial, was addictive.

As luck would have it, I enjoyed a dreamless sleep and woke up five minutes before my alarm sounded. It was a good omen, signaling that not only had I had a perfect amount of sleep, but I was mentally refreshed. Being with Vlad was not for the fainthearted, I knew. It was high time I enjoyed myself, and Vlad had promised entertainment.

The lunch at the Castle promised to be sumptuous. I decided to have a light breakfast, a soft-boiled egg and green tea. I peeked in on Ann. She was on the balcony reading the paper, looking like a lady of leisure. She didn't see me, and I closed the door quietly, making sure it was locked.

At exactly 10:30, I heard a slight beep from the street in front of my duplex. A large, black, shiny Bentley had replaced the dusty gray Chevy Impala. Mike was standing in front of an open door wearing a black chauffeur's uniform, including cap. He didn't seem out of place or uncomfortable, in fact, his bearing was rigidly military. I grabbed my jacket and hurried to the car. As I approached Mike he raised an eyebrow, whispered, "It's rented," turned, and extended an arm to usher me into the back seat where Vlad, wearing a black tuxedo with a ruffled white shirt and black velvet bow tie, was sitting. The interior smelled of freshly oiled wood and rich leather.

"You didn't tell me it was going to be formal," I said.

"Just for me. You're fine," said Vlad, adding, "What's past is prologue."

"I'm not sure what you mean," I said.

"We have been through a lot these last few days. I promised not to use a simulation on you, and I won't budge an inch on that promise."

"I appreciate that. I think I have had all the simulation I can handle for a while. I feel better, knowing that you are actually in a tuxedo, Mike is in a chauffer's uniform, and this car is the real deal. But why all the pomp and circumstance?"

Vlad didn't answer, and at that moment, the car pulled smoothly and powerfully away from the curb and I settled back, closed my eyes and imagined I was riding a magic carpet. Adding to this was the quiet and privacy, for the Bentley had a glass partition separating the passenger compartment.

Vlad didn't interrupt my reverie, but when he sensed it was over, he said, "A number of people at this lunch believe I live a charmed life, and for those, jealousy is a green-eyed monster they have to live with. I know it is extremely unfair for me to use my alien simulations to appear to do magic. A few of these people would like nothing better than to see me fail or at least discover how I do my tricks. You, of course, Donald, could tell them exactly how it's all done, but I'm hoping I have your confidence."

So Vlad was going to perform magic! Naturally—that's what they did at The Magic Castle. I assured him he had my confidence, adding, "Besides, who would believe me?" My attention shifted to Mike, who drove the Bentley like he was made for luxury. I was glad for the glass divide. I don't think it really hit me until then that Mike, who was closer to Vlad than anyone else, perhaps did not know what I knew.

Vlad leaned close, "Donald, I have to ask you a favor. While I am doing my tricks, I have to ask that you to do your best not to react to what you see and hear. After this event is over and we are back in the Bentley, I will be happy to answer all your questions. Can you do that?"

I laughed. "Since I've been trained to make presentations as dry as dust, I think I can make my reactions appear that way. I'll think of the Peter Sellers character in *Being There*. If anyone asks me a probing question, I'll say, 'I like to watch.'"

"That's great. Discretion is the greater part of valor." He was coming up with a lot of English aphorisms, I thought, amused. Perhaps it was part of getting into the mood of the Victorian magical charm of the Castle.

CHAPTER TWENTY-NINE

The Bentley pulled up to the Magic Castle at exactly 11 a.m. I'd learned on their website that it was built in 1908 as a private home for a banker. In 1960 Milt Larson, a writer for *Truth or Consequences*, took over the property, remodeled it, and opened the doors to the Magic Castle three years later. Only magicians could be members, and most of the time it was closed to outsiders, except for weekend shows.

Mike played his chauffeur role to the hilt. He jumped out to open Vlad's door. Vlad stepped out and donned a black top hat, the kind Abraham Lincoln wore. Mike raced around to open my door. The Castle, with its rounded rooms and cone-shaped, pointed roofs, looked whimsical from the outside. Palm trees graced the entrance, and the building was flanked by old-style, spherical street lamps. A man in a gray tuxedo and top hat like Vlad's greeted us at the curb. He looked about 60 years old, with a well-trimmed gray mustache and goatee, wire-rimmed glasses and gray hair. He welcomed Vlad with a paternal smile, extending a hand and making a slight bow.

"This is Bradley Eason, our master of ceremonies," Vlad said to me. "This is my associate, Donald Thomas." Bradley clapped me on the shoulder and his handshake was warm and collegial.

"Mr. Thomas, a pleasure to meet you. I am here at your service. If there is anything I can do for you, anything I can get for you, any question I can answer, please do not hesitate to ask. Also please be reminded that you and Dr. Comsky are our guests at the Academy today."

I was tempted to quip that I liked to watch, but just smiled and thanked him. I walked in behind the two tuxedoed men, throwing a glance back at Mike, who was clearly not invited. I heard Vlad ask Bradley to be sure to send a plate outside for him, and Bradley graciously assented.

The interior of the castle made me think of a turn-of-the-century upscale bordello: Tiffany lamps, chandeliers and oak shelves full of leather-bound books. Bradley Eason pointed out various magical rooms on the way to the dining room, a small room for close-up magic work, and the W.C. Fields bar. Its counter, made of dark wood, was spotlighted from above. As Eason passed, the members abruptly put down their drinks and quietly followed us into the dining room. There was one large table in the center of the room set for fourteen, six on a side and one at each end. Vlad was seated at one end, I next to him, Eason at the far end. The others filled the empty seats. Some of them wore jackets without ties, like me. Most, however, sported suits and ties. One thin, unsmiling man wore a black sweatshirt over a white polo shirt. When all were seated, four waiters in black pants and matching vests, white ruffled shirts and red bow ties, handed out leather-bound menus. As I scanned the menu, my mouth watered. There was slow-roasted prime rib, filet mignon, beef Wellington, rack of lamb, sea bass, Scottish salmon and brined pork chop, and those were just some of the entrées. Reading a menu is a vicarious pleasure, like window shopping. I moved on to the appetizers, salads and side dishes. Vlad, too, was studying the menu even though I was sure he had it memorized. At the other end of the table, Eason's menu was also open. Most of the others had put their menus down, so I assumed at least some of the prolonged studying by my hosts was for my benefit. I relished the elegant offerings for another minute, and then quietly put the menu down. There was a tinkling of glasses and shifting of silverware. This is pretty good, I thought, and I hadn't had anything to eat yet.

The waiters reappeared, offering pieces of crusty, hot, sourdough bread and bottles of red and white wines. When all the glasses were filled, the bottles of white were dunked in ice buckets, and the red wine was decanted into wide-mouth crystal flasks. Eason tapped his water glass with the edge of his knife and stood. The room fell silent.

"Gentlemen, fellow magicians, and guests. For goodness sakes, it seems to have taken forever and a day to get Dr. Comsky to come and grace us with the improbable fiction of his magic. It is high time for him to dazzle us with the unexpected, to puzzle our brains and our magical skills with his mysterious illusions. It is a foregone conclusion that I will be mystified. Gentleman and my fellow magicians, please rise and toast our Dr. Comsky." The room filled with the sound of heavy chairs being pushed back along the tile floor. Everyone stood and shouted, "Here, here!" I rose with the others. Vlad motioned for the group to be seated. "Let's order our food, then I'll tell you what I have in store for you," he said. "Okay?"

At this, the waiters took orders and the room filled with the subtle sounds of bread being buttered, wine consumed, and quiet conversation. The food arrived with a flourish. My Caesar salad was followed by Colorado rack of lamb marinated in garlic and rosemary, with sautéed wild mushrooms, cauliflower, English peas, butternut squash, cured tomatoes, baked potato and creamed spinach. Vlad had ordered an organic kale salad with blackberries, strawberries, raspberries, toasted almonds and blue cheese, with a lemon poppy seed vinaigrette. I asked why he wasn't eating more, and he said, "Not before a performance."

Eventually, Eason rose again, and the room fell into anticipatory silence. This time there was no need for the use of his knife on crystal.

"Gentlemen, fellow magicians and guests. The game is afoot. As you know, I have asked each of you to bring a personal possession that you can hold in your hand. Have you each done that?" Murmuring around the room. Heads bobbing affirmatively. "After this fine lunch, we will adjourn to a special room that you have each inspected. As agreed, the room will have no electronic devices, no recording devices, and you will not be permitted to bring any devices into the room, including, of course, your cell phones. Is that agreeable? And Dr. Comsky would like to offer an additional illusion. Are you up for that?" There was general nodding. "Dr. Comsky? Over to you."

CHAPTER THIRTY

Vlad had adopted a theatrical demeanor, very different from the low-key companion I was used to. It was as if the show was starting, curtains had parted, houselights had dimmed, and mysterious music was playing. I could feel the anticipation in the room. He addressed the group, "Just so you do not think that I am a one-trick pony this afternoon, I plan to attempt a mentalist illusion. Mr. Eason has provided index cards and pens from a local stationery store. He tells me that the cards are still in their original sealed wrappers, as are the ballpoint pens. I have neither seen nor handled the cards or pens. Is that correct, Mr. Eason?"

Eason nodded and held up two sealed boxes. "I purchased them today and they have not been out of my possession. I am happy to have any of you examine the seals, the boxes, the cards and the pens. In fact, if you would like to use your own pen and paper for this illusion, I understand Dr. Comsky will accommodate you. In the meanwhile, I'll pass these around."

All were examined by the magicians on both sides of the table and returned to Eason. Then we followed him down a flight of stairs and through a labyrinth of corridors to a windowless room that held a large, round, clear glass table encircled by fourteen plain wooden chairs. We dropped our cell phones into a bowl at the entrance and found our seats.

The man in the black sweatshirt grumbled. "This is bull, man," he said, but gave up his phone with the rest of us. Eason said, "Now you can go into the room and examine it again to your heart's content, and when you are finished, let me know so that I may enter with Dr. Comsky."

I watched from the doorway. It was interesting to see the magicians inspect the room, some of them tapping on the glass table, some tapping on the walls and some the chairs. A few picked up the chairs, turned them over and shook them. After a quarter of an hour, everyone appeared satisfied. They sat at the table and invited us to join them. We closed the door.

Vlad spoke. "I would like you each to take a card and a pen and write down a word, phrase or sentence that only you know in your own handwriting and then sign your name below it. When you are finished, look up so that I know you are done." Eason opened the boxes of pens and cards, and passed them around the table. Three of the magicians chose to use their own paper and pen. I wrote, THIS IS NOT A DREAM on my index card before I signed it. Vlad continued, "Now fold the cards and the papers, so no one can see what is written. Pass your papers to the person on your left. Take the paper or card you have just received and put it in a pocket that will not be disturbed." When it looked like everyone had complied and I had handed my folded index card to Vlad, he said, "Well done."

Now, gentlemen, please remove the cards and papers from your pockets and hold them in your left hand. Let us start with you, Mr. Eason. Do you remember what you wrote on your card before you signed it?"

"Yes, I do."

"Was it, 'All that glitters is not gold'?"

Eason did not look surprised. "Yes." His neighbor to the left opened Eason's index card, revealing the clearly printed words and the signature below.

Vlad now quickly went from person to person, correctly identifying each of their cards. They were a hodgepodge of the fanciful and mundane: Of course you can read my mind... I'm having trouble concentrating...this is just not right... I should have had the pork chop... Kilroy was here... I want to see more magic...help mother... Can I go home now?... 1235813. "

When it was his turn, the man on my right said, "Don't bother," in resignation. Vlad did not read my card, but just smiled at me and shook his head as if to say, "No, this is not a dream."

Eason started a round of applause and all joined him, some saying, "Here, here!" and "Nicely done." Vlad stood and bowed deeply. He paused dramatically and looked intently at each of us in turn.

CHAPTER THIRTY-ONE

The next thing Vlad said brought a hush to the crowd. "Now, would you each place your personal, handheld item on the table before you." Obediently, each magician placed his item carefully on the clear glass table. A few looked sad, as if it was the last time they would see their precious object and were having second thoughts about bringing it. There was an engraved battered Zippo lighter, a carved piece of jade, an embroidered handkerchief, a small leather-bound book, a tiny painting of a sunflower, a gold coin, a silver belt buckle encrusted with turquoise, a jeweled brooch, a multicolored origami swan, a black and gold poker chip, an old dollar bill, a twenty dollar bill with a corner torn off, and my contribution, a copper penny. Thirteen items in total.

"Gentlemen, please stand and gently push your items to the center of the glass table so that they are all touching. Sit down comfortably in your chairs and bring your chairs close to the table so you have an unobstructed view of your item," Vlad instructed.

"Before I start the illusion I must ask each of you to remain in your seats and specifically not stand up, and even more specifically, not stand and try to touch your item, until I have concluded the illusion. Is that clear? Now we can start. Mr. Eason, is it true that I have never been in this room before?" Eason nodded. "Mr. Eason, is it true that you have provided this glass table and all the chairs, and moreover, that I have had no access to either? Is it true that I have had no access to the objects in the middle of the table?" Eason nodded again. Vlad turned to the rest of us. "Will you all remain in your seats?" There were murmurs of assent. "Now we can proceed. Gentlemen, look very carefully at the objects in the center of the table. Try not to take your eyes off the objects. Try not even to blink."

Gasps erupted around the table. A number of the magicians clutched the armrests of their chairs and started to stand, but forced themselves back.

"What, no, I don't believe it!"

"That's impossible, unbelievable!"

"No!"

And, from Eason, "They melted into thin air."

I looked at the magicians' faces as they moved from side to side in their chairs, sometimes squinting, and some leaning so low that their chins rested on the glass table. They were aghast, appalled and clueless.

I looked at Vlad. He smiled, and I understood exactly what he had meant when he said, I will not budge an inch on that promise. He had kept his promise to not subject me to his alien technology. I could plainly see all the objects at the center of the table. They were still there for me, but for the others, Vlad had created the illusion of invisibility. He had used his Buckyballs to block the visual signal between the neurons of the magicians, so they not only did not see their objects; they saw an empty glass table. They saw the empty glass table because the Buckyballs in their synapses generated that image. After a few minutes the room quieted, and the magicians spoke.

"In my heart of hearts, I have to give the devil his due. Bravo, Dr. Comsky, you have done it again," said the skeptic in the black sweat-shirt.

"I don't believe the naked truth of my eyes, "said a magician with long black hair, black mustache and goatee. "I'm looking, and I don't see."

"This is the stuff that dreams are made of," Eason said. "I cannot imagine how you have produced such an illusion. Well done. Oh, but wait! We have another illusion to experience, don't we? You will bring our precious objects back, won't you?"

Vlad asked if we were all satisfied with the illusion so far, and the nods were vigorous. "As before, keep your eyes on the center of the table and try not to blink." The magicians assumed various pos-tures. While sitting in their chairs, one bent low, and one sat straight, the better to survey the whole area, one leaned over to the right, one to the left, one tried to rotate his head sideways. Others simply relaxed, as if the illusion was so incomprehensible they knew they'd never figure out how it was done.

I knew from their expressions the instant the items were restored to their sight. The moment was just as awe-inspiring as when the objects had vanished. After a few thrilling moments, Vlad said, "You may now stand up and retrieve your objects. Please confirm that they are your objects and have not been tampered with in any way." Each item was inspected, shaken, held up to the light and even smelled to verify its reality. After a few moments, there was a con-sensus that the original items had been returned safely.

As the magicians retired to the bar to have one for the road, they were still talking about their vanishing objects. They said that the mental trick could have been done several ways; whereas they had never seen objects vanish without smoke and mirrors, right before their eyes. Vlad thanked Bradley Eason for his hospitality and we excused ourselves. I had seen the looks of envy in the eyes of all, and covetousness, too. Vlad had a secret and they wanted it, but no one had it but me.

CHAPTER THIRTY-TWO

Mike and the gleaming Bentley were waiting out front. As he opened my door he asked how the performance had gone.

"They all seemed to be astonished," I said.

"Guess you'll be invited back for free dinner?"

"I wouldn't be surprised," I said. He closed my door and hurried around to let Vlad in. As we pulled away, Vlad made sure the window closing off the driver was closed and asked if I had any questions. I shook my head. I said I thought it must be easy to do magic when you have an astonishing technology. We were quiet for a few minutes as I let my mind drift. I thought of being at home, perhaps looking in on the miraculously cured Ann. "Can I ask you to do a favor for me?" I said.

"You can ask," he said without hesitation.

"Could you give Ann a dream, a vision, a memory whatever you want to call it, so that when she wakes up tomorrow morning, she will still remember it?"

"I could," Vlad said. "What did you have in mind?"

"She told me that some of the happiest times in her life were when her husband, Richard, was courting her. Do you think you could do something to get her to relive or re-experience or just dream about that happy time?"

"That seems like a harmless enough request. Let me see what I can do." The Bentley floated north on Highway 5. The forest of Griffith Park loomed as a great dark shadow on our left. After a while it was Vlad who broke the silence.

"Donald, you have had a lot of excitement this evening. Are you tired?"

"Actually, I'm quite awake."

"Want to witness something I don't usually do?

"What do you mean, Vlad?" He seemed to be unusually hesitant and circumspect.

"I don't usually like to interfere in people's lives in an overt way," he said slowly.

I couldn't help replying, "You mean like with me?"

He smiled wryly. "Not quite. As you may remember, you gave me your permission." My heart leaped. He was talking about something new—my curiosity was mixed with anxiety. Who could not be curious about an alien? Or worry about what he might conjure next?

I took a deep breath. "If you're asking for my permission to be in your simulation, you have it."

CLICK

CHAPTER THIRTY-THREE

This simulation was different, very different, from everything I'd experienced before. I felt I was a disembodied spirit, I could hear Vlad, but I could not see him. It was like watching a movie, except I was wearing virtual reality goggles so that I could move my head and eyes around to see 360 degrees,

Vlad asked, "Are you all right Donald? Disoriented? Feeling queasy?"

"No, this is great. It's like being in a 3-D movie." And in the next moment we are taking a ride into space. It was like a view from Google Earth. It was nighttime and we were looking straight down on the city of Covina. The San Gabriel Mountains were a jagged shadow in front of us and the City of Industry was a blinking mass behind us. As if in a helicopter, we swooped down past Highway 10, headed towards the 210, then lowered almost to street-level. I could see by the street signs that we were on North Glendora Avenue, headed north. The streets were empty, except for one white Cadillac STS. There were three young men in the car. I recognized them. They were "Marvin Gaye" with his brothers Hank and William. Marvin was driving; Hank was in front and William in the back seat behind him. I could hear their conversation.

Hank said, “So, Pauly, my man, why are we going to San Dimas again? And why are you driving on the streets?”

“I told you before, they've got some good low-price coke that we can cut and make a whole lot of money. Don't you get it? Drive a few extra miles and make a lot more money?”

“And why are we on the streets again? And what about those hundred dollar bills you won't let us spend?” William asked from the back. I thought, at least I now know one of their real names.

Half-turning to the back, Pauly answered sarcastically, “Don't you know there are lots of cops on the freeway at this hour? All looking for drunks. And just shut up about those hundred dollar bills. I told you I had a dream. I told you to think about it like lucky money.”

“Yeah, okay, we get it,” Hank mumbled to himself.

They rode quietly for a few moments. Then a bright white light filled the interior of the Cadillac. Pauly jammed his foot on the brake and the car stopped at East Clover Place.

“What's going on? “ Said William.

“What's that light?” Said Hank.

Pauly opened the car door and pressed his hands to his eyes, moaning, “I'm blind. Everything is gone black.” Lurching from the car, he fell to the asphalt. Then we heard a loud voice, the kind you'd hear in a baseball stadium when the announcer is shouting about a triple play.

“PAULY, YOU ARE NOW BLIND. IF YOU GO TO SAN DIMAS IN THREE DAYS AND ASK TO HELP THE CHILDREN, I WILL RESTORE YOUR SIGHT.” I recognized Vlad's voice.

Hank and William, both rubbing their eyes, rushed to help their brother. Hank asked, “You all right, man? What happened?”

“I don't know but I can't see,” Pauly moaned.

“As soon as that light went away, we could see fine,” William said.

"And who was that talking to you?" Hank said. "That was creepy, man." The two stared down at the stricken man, who lay, blind, on the black and cold pavement. "Sheeeet!" Hank and William sighed together.

CLICK

We were seated in the Bentley, gliding to a gentle stop in front of my duplex.

"Did all that just happen for real?" I said.

"Yes. Are you all right with that?" Vlad asked softly.

"You are dangerous, and I like it."

I got out of the car and the Bentley disappeared into the night. And what a night it had been. I looked in on Ann and wondered what kind of good dreams she would be having. Neither she nor the cat stopped snoring when I peeked into her room. I left without waking them.

CHAPTER THIRTY-FOUR

The next day I made two phone calls, the first to Pebbles, to see if I was missed. He suggested I could stay away for a few weeks. Everything was rolling along as anticipated. "It's all good news," Pebbles said. "The electrodes are functioning according to specifications, and I'm considering an installation in the motor cortex within a month. The arm is performing well when hooked up to the lab computer."

Typical Pebbles—quick and to the point, like one of his quarter horses.

The second call was to Rita Goldblatt. She asked how I was and what I was up to. I told her about all the good work that the medical military community was doing for its injured veterans.

There was a pause. "Did the card work?" she asked.

"I didn't get a chance to use it," I lied.

"Have you given up on lucid dreaming?"

"For awhile."

"Oh." She sounded disappointed. Then she said what she always said every time we parted: "You know Donald, you can call me anytime."

I would have liked to explain to her that there was a new type of dreaming, one that was brought on by mechanical means. I could tell her that it was just like having a TV set in your brain and a remote control to change channels. That it was easy. I knew that Vlad wanted me to keep his alien identity a secret. I had his confidence, and I would respect it.

I drove to Venice Beach. It was still the cool of the morning when I arrived. I walked on the sand and stood in the place where Vlad and I had sat in our simulated beach chairs. I wanted to regain my perspective. I had been here with an alien in his simulated world.

I sat on the sand reading the Los Angeles Times, letting my mind drift over the news of the day. City politicians were asking the federal government for money to deal with congestion on the freeways. The temperature in Los Angeles continued to rise, but there was good news about the decrease in the smog level. On the bottom of the third page a small headline caught my eye. POLICE ARREST MURDER SUSPECT. At 1 a.m., neighbors had responded to a scream in the apartment of Jacqueline LeRoux, 67, in the 300 block of Crenshaw Boulevard in Los Angeles. Police called to the scene arrested William Strakel as he tried to leave the scene. Evidence was found in Strakel's vehicle linking him to other burglaries and to the unsolved Los Angeles Strangler murders. The story went on, but I had to wonder if Vlad had anything to do with this, or if he had just let nature take its course.

Perhaps I will never know. Perhaps I will never want to know. In any case, I was glad I had let nature take its course, at least for a time. I lowered the paper and looked around me. The real Venice Beach was beautiful. It was beautiful seeing real people. Had the idea of being in a simulation lost its novelty? What would Pauly, Hank and William be thinking now? Would the idea of a simulation even cross their minds? Had the idea of having an extraterrestrial as a friend lost its novelty? I thought not!

CHAPTER THIRTY-FIVE

Ann was very happy to see me when I arrived at the duplex. She reached out and pulled my face toward hers for a big kiss on my cheek and a warm hug.

"It's so good to see you, Donald. I have something to tell you. Please sit down with me."

"Of course, what is it?"

She was excited, and her words tumbled out. Her face was pink and her eyes fixed on mine with excited seriousness.

"Last night I had a dream," she said. "It was about Richard. He was young, handsome, well-dressed. Like he was during the months we spent together before we were married. We were so happy then! I can't remember it all now but when I woke up, I didn't want the dream to end. I wanted to be with Richard again. I miss him so much."

"Of course you do," I said. "You look so happy." Vlad had exceeded my expectations.

"Will you let me cook for you?"

I grinned widely. "How can I refuse? Would you mind if I invited your doctor?"

“You mean that miracle worker, Dr. Comsky?”

“Let me call him and see if he's available.”

“You mean, just like that you expect him to drop everything? Oh, Donald, he won’t be available.” I could see the hope in her eyes.

“I can ask,” I said, knowing that Vlad’s time was different from our time. He seemed to make time expand, and make himself completely available to me. Come to think of it, that is a stupendous quality in a friend. I took out my cell phone, pushed Vlad’s autodial number. He answered on the first ring. I extended the invitation and he immediately accepted.

“Yes, Donald, what time? In addition, please ask Ann if she would mind if Mike came. Tell her he's a good eater.”

I asked Ann, who was seated at the kitchen table, already turning the pages of a recipe book.

“Of course he can bring a friend, the more the merrier. Seven?” She answered, beaming.

“See you at seven.” Vlad said into the phone.

CHAPTER THIRTY-SIX

Ann's doorbell rang promptly at seven and I answered the door to find Vlad and Mike very neatly dressed in slacks and sports jackets. Vlad was holding a dozen red roses in one hand and a large, flat package wrapped in plain brown paper in the other. Mike held a two-pound box of See's chocolates and a bottle of Italian Chianti.

"I love the smell of garlic," Vlad said. I almost asked if there was garlic on Paucri, but stopped myself in time.

"It's garlic bread," Ann said, pleased. While she put the final flourishes on her angel hair pasta with marinara sauce, Mike arranged the roses in a vase. Over dinner, Vlad engaged Ann in conversation about her childhood in Milan, Italy. She was impressed that he knew so much about her hometown and could speak Milanese with only a slight accent. He was charming and attentive but it was Mike who impressed her as a great eater.

The food was delicious. Ann had added fresh mozzarella cheese to the cooking sauce and let it melt before adding the pasta. It was a wonderful combination, especially with the Chianti. At the end of the meal, as the lively talk tapered off, Vlad removed the brown wrapping from his package. "This is a gift to you, Donald," he said.

It was a painting, blue and white and square. I looked at it dubiously," What is it supposed to be? Abstract art?"

"You don't like it?" Vlad asked, looking crestfallen. "It's a marker. It reads one hour and twenty minutes on the galactic clock, or 36 thousand years from the time the first true circle was drawn on a French cave wall, to the time the simple perfect square was drawn. At one hour and twenty minutes, this makes Earth a seven on the Cosmic Development Logarithmic Scale, in other words, a major event. As a comparison, a six on the CDLS would be thirteen galactic hours and twenty minutes, or 360 thousand Earth years and a strong event."

I stammered, "But Vlad, what does this mean?" I looked at the painting, trying to take in all that Vlad was saying. This object was a mile marker on a galactic scale. How was that possible? And then I got it. This milestone measured the rate of human development on a galactic scale.

"So to be clear, Vlad, this is the equivalent to a seven on the Richter earthquake scale?"

Ann and Mike listened avidly but without understanding what we were talking about. In fact, I suspect that Vlad had done something so that they weren't even aware of the conversation between me and Vlad.

Vlad smiled a radiant smile. "Yes, and the San Francisco earthquake of 1906 was seven point eight. Human development is moving along at an astounding rate."

Without thinking, I asked, "Is that a good thing?"

"Well, it is not a common occurrence." This was the first time I had seen the painting that was copied to Scientific American. It was the mathematical answer to the question, what is the least number of squares of different sizes that could fit into a single square? The 21 individual squares were each painted white on one side of their diagonal and blue on the other.

Ann excused herself from the table, saying, "Dr. Comsky, I don't know what you are talking about, but Donald looks pleased with your gift, and I thank you and Mike for the roses and candy. Now I have something for you."

She returned with a folded sheet of paper. Handing it to me, she said, "This poem was originally written in Spanish about a real incident experienced by the poet. It was passed around the Spanish senior centers here in Los Angeles, and a translation made it to mine." She looked tenderly at Vlad. "I was touched by it. It made me think of the first time I saw you, Dr. Comsky. I was in such pain, I had lost hope and you came into my life. You took my pain away; you took my disease away. Yes, I know you are a doctor. It is your job. It is what you do. But you did not have to do this for me. You went out of your way and changed my life." She had tears in her eyes as she extended a trembling hand holding the poem—trembling not from disease but from emotion. "Please read the poem to us, Donald." I began reading.

I was sleeping and YOU awakened me,
I was dirty and YOU washed me,
I was hungry and YOU fed me,
I was lost and YOU reached out to me,
I was a stranger and YOU gave me friends,
I was beaten and YOU comforted me,
I was damaged and YOU cleaned my wounds,
I was a drug addict and YOU gave me hope,
I was dying and YOU saved my life,
YOU came to me as in a dream,
The sunlight poured from behind YOUR head,
I could not see YOUR face,
Yet YOU spoke to me as my neighbor,
YOU spoke to me as my friend,
That such a one as YOU exists,

In this world today,
Is a gift I find hard to believe,
YOU have restored my faith and,
changed my life till its end,
and I do not know YOUR name,
my Samaritan.

The note at the bottom of the poem read, "Written with all humanly possible humility and gratitude, Juan Carlos Menchu."

During the reading I saw that Ann's cat had jumped onto Mike's lap. Mike didn't seem to mind and continued eating big mouthfuls of spaghetti, twirling it with his fork and soup spoon. The cat instantly fell asleep and started snoring. Mike started to reach for his napkin but caught himself, not wanting to disturb his snoring guest. He wiped the corners of his mouth with his thumb and index finger and said, "That was beautiful. The world would be a better place if there were more people like that." The sudden memory of Juan Carlos Menchu's near-gangrenous foot momentarily overwhelmed me. I didn't know if Mike meant Juan Carlos or the Samaritan. I doubted if he remembered Juan Carlos. There were probably many such men in his and Vlad's life of service, and in either case, I agreed with him. I glanced at the painting propped up on a chair behind Vlad, a painting that was a cosmic mile marker. Was this Vlad's view of time? Did all of human development, to him, just happen in an hour and twenty minutes?

It was a strange world I found myself in now, sitting at dinner with my longtime friend, Ann, as happy as she'd ever been, with my new, benignly lethal friend Mike, and with someone who had achieved his goal and made contact with an alien. That someone was Vladimir Comsky, and I was the alien. I had so many questions I wanted to ask, but this was only the beginning of a strange and wonderful relationship.

I'm grateful that Vlad convinced me to write this story from my perspective. So much happened in such a short time that I barely had time to take a deep breath. Writing has provided both time and breath. Vlad came into my life like a meteor. He wiped my former life away. The light he gave me allowed me to see beyond myself, just as the light he blinded Pauly with has, I'm sure, changed his life. I'm still not sure why Vlad chose me, but if I have made my experience in meeting him understandable, I have achieved my, and perhaps his, purpose.

I'm back on that roller coaster at Coney Island, poised in a moment of stillness right before the world drops away.

THE END

Made in United States
Troutdale, OR
11/03/2026

56844885R00086